THE MUSCLE BOOK

Anthony Serafini

Arco Publishing, Inc.
New York

Published by Arco Publishing, Inc.
219 Park Avenue South, New York, N.Y. 10003

Library of Congress Cataloging in Publication Data

Serafini, Anthony.
  The muscle book.

  Bibliography: p. 138.
  Includes index.
  1. Athletics. 2. Muscles. 3. Strong men.
  I. Title.
GV706.8.S47   796.4'1   81-4654

ISBN 0-668-05088-8 (Library Edition) AACR2
ISBN 0-668-05092-6 (Paper Edition)

Printed in the United States of America

10 9 8 7 6 5 4 3 2 1

Contents

Acknowledgments

I would like to thank the following people and organizations for their generous assistance in supplying information and photos for the book:

Kathy Tuite of the Dept. of Audio-Visual Productions of Purdue University; David Webster; Doris Barrilleaux of the *Superior Physique Association*; Ben Weider of the International Federation of Bodybuilders; Franco Columbu; Bob Klez; the *United States Sports Academy*; Peter Griffin of *Sport* magazine; Jeffrey Thrift, distinguished track & field official of Great Britain; John Grimek of *Muscular Development* magazine; Bob Kennedy of *Muscle Magazine International*; Don Arbour of the Department of Athletics at Princeton University; Prof. G. Kenntner; O.I. Solf of *Deutsche Sportbund*; Dr. Horst Ueberhorst; Chris Dawson of the Sports Information Department of the University of California; Prof. Al Thomas of Kutztown State College; the Finland National Tourist Office; Vic Kelley of the University of California at Los Angeles, Dept. of Athletics; Ron Dinkel of the Sports Information Office of Emporia State University; Linda Gray of the Bulgarian Union for Physical Culture & Sports; photographer Stan Pantovic; Orin J. Heller; Mike Lambert of *Powerlifting USA*; the Embassies of Japan, the German Democratic Republic, the Federal Republic of Germany; the Austrian Embassy, the Embassy of Spain, and the Embassy of the USSR; *Women's Sports* magazine; CBS-TV for photos of the "World's Strongest Man" Competition; the AAU; Ron Paradis, Director of Intercollegiate Athletics at the University of Oregon; Mabel Rader of *Iron Man* magazine; Ed Corney; Pete Cava of the *Athletics Congress of the USA*; the Dept. of Athletics of SMU; Dick Quinn of the US Sports Academy; Warren Morgan, photographer, of the *Eugene Register*; *Track & Field News*; Tufts University; Kathy Slattery of the Department of Athletics of Dartmouth College; the *Paul Anderson Youth Home*; the Canadian Consulate & the Public Archives of Canada; L.B. Baker of the *World's Pro Wristwrestling Association*; Dept. of Athletics of Lehigh University; Jill Ganger, the United States Embassy in Moscow; San Jose State College; David Willoughby; Pat Malone of Purdue University; Jan Dellinger of *Strength & Health* magazine; Hannes Strohmeyer of the *Institut fur Sportwissenschaften der Universitat Wien* & Clyde Emrich, strength coach of the Chicago Bears.

I want to give special thanks to the Department of Philosophy of Princeton University for appointing me Visiting Fellow for the Spring term, 1980, thus giving me the time needed for my research. My gratitude extends as well to Stephanie Lewis of the department, for fine conversations on the philosophy of muscle. My wife, Tina Potecha Serafini, was, of course, a *sine qua non* of the entire project. Her enormous patience, tact, and trenchant observations in editing the manuscript created a delightful atmosphere.

Introduction

My mania for muscle has existed from as far back as I can remember. In those days all was well with the world. Ike was in the White House, the Dodgers were in Brooklyn where they belonged, and Steve Reeves was the king of Muscledom. Yet, I've always felt that muscle-building—this basic, primordial element of humanity—has too long been neglected in the sporting literature. To some extent this situation has improved with the publication of the now classic *Pumping Iron*. But bodybuilding is only a thin slice of a much larger pie. As long as man has had to fight nature with sheer strength of limb, there have been games to celebrate that fight. Besides the popular *hammer toss, discus,* and *shotput,* besides the now-popular *bodybuilding, olympic weightlifting,* and *powerlifting,* there have existed for untold millenia the unheralded but equally demanding and competitive muscle sports of many lands and many cultures.

It is, to my mind, astonishing that with the dominant role of muscle power in so many sporting activities, these sports in all their glory and misery have never been collectively described. This book is an attempt to remedy that sin. It isn't an instructional book. The guiding principle is to explain, portray and convey some of the sense of wonder and awe that I (and I believe many others) have found in the apocalyptic struggle between force of Muscle and force of Nature. Only those sports where sheer physical power is a dominant ingredient are found here. Besides the better-known sports mentioned above, I've included such exotic fare as Austrian *stone-lifting,* Scottish *caber-tossing,* and Basque *stone-lifting.* To understand these sports is, in some sense, to grasp the history and the struggles of their peoples. For, as with so many areas of existence, the history of a nation's sport mirrors the history of the nation. To this end, I have tried to convey some sense of the cultural, social, and philosophical roots of these many forms of sporting life.

Along the way I have tried to dispel some of the most long-standing myths about strength and strength athletes. What of such shibboleths as "homosexuality," "musclebound," and the legend that women are "weaker" than men? To those still enamored of the latter notion, for example, I invite a perusal of the section of women's powerlifting and the exploits of Jan Todd. And, as a callous youth racing full-throttle toward middle age, I've tried to dispel the notion that strength vanishes with time. How many would dare to match physiques with 57-year-old Roy Hillegen, or 65-year-old Vince Gironda? Would any but the most foolish venture into the strength arena against the 46-year-old Paul Anderson? Who is there to try besting the legendary George Clark (now 79) in a one-on-one toss of the awesome *Braemar Caber* of Scotland? And a king's ransom to those who would lock wrists with 80-year-old *Bob Hoffman* of York, Pennsylvania!

Undoubtedly, the book will not satisfy all. Some will argue I might have gone deeper into the philosophical significance of these great sports. Others will argue against the inclusion of wrestling because of the importance of technique in that sport. Still others will inveigh against the absence of football. To these I can only say that I've made the selections on the basis of the logic of the sports themselves, the opinions of experts, and my own lifelong battle for gigantic muscles. Though football professionals both develop and use enormous power, football is not, by my reckoning, a strength sport because of the overriding role of strategy and tactics.

The saga begins with a seamy first-person account of my own joys in the exotic and wonderful world of bodybuilding. I treat the subject this way because bodybuilding has, in my view, a rather special place in the muscle world. The one sport where muscles are developed for sheer aesthetics rather than brute strength, it is perhaps the most essentially personal and recalcitrant to objective description. Perhaps the nearly unreachable combination of power and perfection of physical form is the ultimate goal of every strength athlete. For this reason I begin the book with this sport, art, or insanity—as you please. The other sports I've selected for inclusion represent, I believe, virtually every conceivable organized and institutionalized strength sport in the world today.

By better understanding the role of strength in the human psyche, we may better understand ourselves. It has become fashionable in recent years to philosophize interminably about the nature of the soul. Psychologists try to measure it, my colleagues in philosophy try to define it, and the "Moonies" try to "save" it. Yet they may all have overlooked another axiom dating back to the Greeks; according to Aristotle's *De Anima*, Man is essentially a union of both mind *and* body—we cannot understand the one except in terms of the other. Far from an abstraction to them, the Greeks put it into practice in their early Olympic games. Perhaps a look at these sports and games will give more insight into the nature of the human soul then an academic text ever could.

At the very least, I hope it will give some small degree of pleasure.

1

Bodybuilding:
The Art of Muscle

THEORY

I am not Arnold Schwarzenegger. I am Anthony Serafini. There are differences between us. The "Austrian Oak" has a 22-inch biceps and a 52-inch chest and has not, to my knowledge, ever studied the later works of John Scotus Erigena. Also, he has been pumping iron longer than I have. There are, of course, similarities. We have, manifestly, identical initials. I am certain neither of us has ever failed to open one of those damn 7-Up screw-up caps, and he probably possesses my awesome capacity for opening aspirin safety bottles without lining up the arrows. I have evidence that he even plays the violin with, to be sure, considerably less virtuosity than Joseph Silverstein, Pinchas Zukerman, or even me. Sadly, however, my biceps is still well under the 22-inch mark. But there is hope. Every bench press I perform crushes the life out of another muscle cell, which will, like Lazarus from the dead, rebound in Olympian proportions. At least that's the theory.

The arena in which this metamorphosis takes place is called Mike's Gym. Mike's is not just any gym, you understand, but a world-renowned, hard-core bodybuilding emporium: noisy, sweaty, and supersaturated with muscle and iron. In appearance it is mercilessly utilitarian. (You can't build muscle in "pretty" gyms.) All great gyms, like Gold's in California and Mid-City in Manhattan, are like this. Places like these are to be carefully distinguished from "spas," "health clubs", and the like. They are beautiful in a way, and very moral, unlike many of the "spas." There are no overweight businessmen and no "disco body shapers" there. Nobody pressures you to join for two millenia, and there are no attendants with lily-white uniforms and ear-to-ear smiles.

1

The author pumping iron. (Tina Serafini)

Mike's is located in Cambridge, Massachusetts, on the top floor of a factory. Getting in is a nontrivial task. You must first run the gauntlet of diminutive Mediterranean typesetters who scowl menacingly at you as you thread your way through a labyrinth of machines and puddles of ominous liquids. Once on the gym floor you feel like Gulliver among the big guys. In the far corner stands a massive Vietnam vet playing with a 475-pound barbell. He doesn't think about contests because he's doing it "just to keep in shape." In the center an arm-wrestling champion is doing lat-machine pulldowns with, apparently, all the iron the machine will hold. Off to the right a team of powerlifters do alternating quarter-squats with 1000 pounds.

The most impressive and terrifying aspect of Mike's Gym is the machinery. "Nautilus" machines: The mere mention of them ignites controversy in all serious bodybuilding circles. They are unbelievable machines. In the center of the gym is an enormous contraption that resembles the offspring of some unholy union between an industrial forklift and an electric chair. The general idea is to strap yourself in (this prevents your spine from snapping), load up the platform with as much weight as you can handle, and try to haul it up as far from the floor as possible. This develops the latissimus muscles and the upper pectoral region. Closer to the wall is the triceps-extension machine. The idea here is to lock your arms behind two padded, weighted plates and extend them to full length above your head. This develops the triceps. This particular machine is feared by even the mighty. My theory is that should you get stuck in the "down" position, you have two options: You can either be stuck there, frozen into the position of Luis Tiant winding up to throw a changeup, or you can brace your legs against the opposite wall, walk up until you can hook your feet over the overhead lamp, and then do a sort of leg "pull-up" till you're out. Your task then is merely to get off the lamp.

Once pried out of the machines, one goes to the barbells. These, of course, are the main diet of the bodybuilder. No one, not even the most devout Nautilus enthusiast, would try to develop his physique without them. For developing the legs, many prefer something called the Jefferson lift. The procedure here is singularly unaesthetic. It involves straddling a heavy bar and grasping tightly, proceeding to bob up and down in the manner of Falstaff pulling up an ill-fitting pair of pants. The chief argument for the machines is their supposed safety compared to barbells. But Griff, the gym supervisor, is careful to the point of paranoia about the participants lifting carefully and with correct form. Frequently, an overenthusiastic barbell lifter will attempt an overhead lift beyond what he can safely do. On such occasions balancing the bar becomes a real problem. Should one arm begin to sag, the weights can very quickly slide off the ends. In short order, a weight room can be transformed into an artillery range of five- and ten-pound projectiles. A barrage of such, whizzing every which way, can turn the most macho of bodybuilders to jelly. This has never happened at Mike's.

Bob Klez, Junior Mr. America, working out. (Anthony Serafini)

One of the great myths about places such as Mike's is the assumption that there are weirdos of all ilks lurking around every corner. The closest I ever came to such was when Joe, the Marine veteran, asked me to jump on his back so he could do something called a "donkey calf-raise." I agreed, fearing the consequences of a refusal. Later somebody told me that it was OK because Schwarzenegger regularly did that exercise with three guys sitting on his back and a girl doing something else.

For the casual onlooker, one of the most curious aspects of Mike's are the mirrors placed at strategic places around the floor. To the untutored mind this signifies a peculiar kind of narcissism. It is not. The bodybuilder considers himself an artist. He carves in flesh with as much thought and sensitivity as Polyclitus did in marble. Just as the sculptor constantly observes and refines his product, so the bodybuilder must observe, carve, and finish the body. And this can be done only with mirrors. Frank Zane, currently Mr. Olympia and one of the world's most symmetrical bodybuilders, achieved his goal by paying constant attention to the equal development of all three heads of the triceps and to the proper proportion of forearms to upper arms. This would have been virtually impossible without mirrors. For the fledgling bodybuilder like myself, mirrors do not have top priority as training equipment. A rough outline is the order of the day. Details will come later.

Contrary to popular opinion, the whole business is very scientific. The aspiring bodybuilder follows a diet designed for gaining as much muscular body weight as possible. This process is called "bulking." Then he goes on a high-protein, low-carbohydrate diet. The point is to tear away all vestiges of body fat, leaving nothing but muscle—each muscle being clearly set off from the others. Bodybuilders call this *definition*. The best bodybuilders have an elusive quality called *symmetry*. A symmetrical bodybuilder has every part of his body in proportion to every other. Not many people can build symmetry. Most have some structural or other deficiency that even exercise cannot correct. There is not too much one can do about this.

A bodybuilding workout has been likened to work. It is, quite clearly, nothing of the kind. Schwarzenegger compared the feeling of a good "pump" (filling the muscle with oxygenated blood) to orgasm. That is a partial analogy at best. Some people consider it a mystical experience, but this too is only part of the picture. It is simultaneously a satisfying, invigorating, transcendent, and overpowering fantasy trip. One gets the feeling, blasphemous though it may be, of having been one with the divine (unless, of course, you get stuck in the triceps machine).

Closer to reality, being a musclehead can make the more humdrum aspects of social and professional existence a nightmare; but you fight back. For me, academic cocktail parties are becoming a problem; especially philosophical ones, where the only muscles one finds are in a cream sauce.

"How do you do, sir?" asks a Harvard graduate assistant. "What is your area of specialization?"

"Well, I'm a musclehead and I'm working on my triceps right now," I answer laconically.

"Oh, well, ah . . . I see . . . isn't that interesting!" Now the cross-examination begins.

"Is it true that bodybuilding makes you muscle-bound?" (Ah, well, at least they don't ask me my philosophy of life here!)

"Well, yes, definitely, definitely. In fact, I've been noticing that my octaves and double-stops in the first movement of the Mendelssohn Violin Concerto don't seem to be going as smoothly lately. On the other hand, the flying staccati and two-octave chromatic runs in Paganini's Caprice No. 5 seem a bit easier, mostly because I added an extra set of dumbbell curls to my workouts. So there are pros and cons. Would you like me to demonstrate?" By the time I have the bow tightened and resined, I am resolutely urged by my inquisitor that further elaboration is unnecessary. I play a few arpeggios anyway, just to pound the point home.

Next question: "These musclemen, are they, well . . . smart?"

Detecting the nonverbal concession that I must be some strange anomaly in the muscle world (since we both studied the essence of Descarte's navel under Professor Schlump at State U.), I respond: "Well, probably not, since the national junior weightlifting champion, whom I knew at Cornell, never made it to Harvard Med and finally had to settle for Stanford, poor fellow. But wait till I tell you about what happened to the rest of the weightlifitng team!" Strangely, they're not interested.

"Incidentally, how about a game of chess?" they sometimes ask.

Oh, but I am a man without pity. "No, thanks. I haven't played much since I beat Grandmaster Bent Larsen in an exhibition game some years ago" (A bit misleading—I never, ever, tell anyone he was playing 45 other people at the same time).

Of course the hard questions come, inevitably, and they are always the same. "Why do you do this to yourself? Do you consider it an art or a sport?" To these, I have no easy answers. The bodybuilder has, to be sure, a colossal ego. His body is his life, his anima. These things are the same in any field; The Arnold Schwarzenegger of the posing dais is the Itzhak Perlman of Symphony Hall, the Bobby Fischer of Reykjavik—any master of a craft. The motivation, the reward, and the frustration—they too are all the same. Louis Ferrigno would have found a kindred spirit in Boris Spassky.

As to whether bodybuilding is a sport or an art, my belief is that the question was answered long ago in a different context by the great chessmaster Emanuel Lasker. In an interview in St. Petersburg in 1914, just prior to his famous victory over Capablanca, Lasker replied that chess was neither wholly art nor wholly sport, but a "struggle." These remarks apply to bodybuilding. It is a struggle to physically represent an abstraction, the concept of perfection. It is an attempt at conquering matter. To unite form and substance is the desire of the bodybuilder. In our monumental arrogance, dare we think that bodybuilding transcends mere imitation? That Schwarzenegger transcends Michelangelo, doing in flesh what the master managed in the much easier medium of stone? In a certain sense the difference between the body of Schwarzenegger and the statue of David is the difference between Aristotle and Plato: the difference between the abstract universal and the concrete particular. In *Pumping Iron*, Charles Gaines states it simply: "The bodybuilder is an idea made fact."

What of the product itself? Granting the psychological and aesthetic motives of the bodybuilder, why should the average person regard the body as a genuine object of art? Is it just a tacky rationalization to argue this way, or are we freakish monstrosities, as the common wisdom would have it? Can the argument be made that how Arnold looks is how people should look all the time? I think so. The human body is, after all, a functioning, goal-oriented entity. The muscles are designed to work, to accomplish a task. The metaphysics here are not new. It is a teleological question. According to Aristotle's *Metaphysics*, all things in nature are designed for, or aim at, some end. Striving, growing, and change are the essence of all that is. The end that he speaks of is embodied in all things. Just as the acorn is predetermined to strive for its destined end, so all human nature strives for some ultimate ideal. The determination of what the "end" of all things and processes is, is a matter of observation. Aristotle insists that what people ought to strive for is simply what they do strive for; what the acorn should become is simply what it does

Bodybuilders regard the body as a genuine object of art. (Anthony Serafini)

become. There is no dichotomy between facts and the values in Aristotle. The bodybuilder is therefore, at least subliminally, an Aristotelian. He insists that this is how the body should look, merely because this is how the body does look when pushed toward its end, its outer limits. The logos of muscle is work—pushing and pulling against resistance. That is its function. It does not, therefore, require extraordinary powers of deduction to conclude that if spending several years pumping iron causes the muscles to look like Arnold Schwarzenegger's, then this is how they ought to look. Aesthetic preference for the superhuman physique is therefore not a matter of taste, but a matter of logic. The person who finds Arnold Schwarzenegger "overdeveloped" is displaying not merely bad taste, but bad logic and an ignorance of the process-oriented nature of the physical universe.

Sadly, even the most sagacious arguments are lost on many, especially at academic functions. In the last resort, one can always cite more practical applications of the craft. A particular favorite of mine consists of rolling up my sleeves every time students ask for extensions on term papers. (Their grandmothers are sick. Their grandmothers are always sick.) The very sight of all that rippling bulk makes them cringe with terror. Thus do I halt the deterioration of American higher education.

PRACTICE

But what, on a more practical level, is bodybuilding? I claim it is art, some call it sport, and others argue that it is the most healthful, vital, and rejuvenating exercise in which man has ever engaged. In its purest state (the form manifested by the greats, such as Steve Reeves, Franco Columbu, and Arnold Schwarzenegger) it is the quintessential expression of form—the development of the body for development's sake.

Yet it is also a practical art. The practical bodybuilder has two goals: health and ultimate physical development. Its benefits are documented, wide-ranging, and in many ways unique. Despite the vogue today for jogging, swimming and tennis, these activities can't compare to bodybuilding in overall health benefits. The celebrated advantages of these sports to your heart and wind are at least equalled by bodybuilding, without the damaging side-effects associated with the former, while offering unsurpassed benefits to normally weak, under-exercised muscles.

To the bodybuilder, considerations of raw strength are secondary to general physical development. This distinction has historical roots. Man has always practiced strength development. The use of progressive resistance against gravity to build strength goes back as far as the ancient Greek wrestler Milo, who is reputed to have developed enormous strength by carrying a baby calf on his back every day until the calf was fully grown. Yet the history of strength for strength's sake is the history of what is now called weightlifting, not bodybuilding. The idea in weightlifting is to lift as much weight as possible from the floor to arms' length overhead at one time (or up to waist level only, as in powerlifting).

It is also a great sport, with many great practitioners, some of whom (such as Franco Columbu and John Grimek) have doubled as bodybuilders. Nevertheless, Olympic weightlifting is not bodybuilding: aesthetic considerations are secondary in the former (witness the great Russian lifter, Vasili Alexeev). Lifting the weight is the one and only object in that sport.

In recent years, however, men noticed that along with the awesome strength that develops via progressive weightlifting, there came a hitherto unnoticed fringe benefit: a superhuman physique that causes the ordinary man to cower as if in the presence of some Altantean god. It was discovered that the body can be an object of art as well as an engine of power. Bodybuilding was born.

Unfortunately for the sport, it became almost immediately the object of derision. It was suggested that bodybuilding contests were seedy events conducted in damp basements, attended only by winos, voyeurs, and fringe characters of suspect sexual preferences. Musclebound became the rallying cry for untold numbers of parents who worried that pumping iron would result in an epidemic of teenage heart attacks. It was alleged that women were repulsed by the sight of such physical development (yet they fill the first ten rows of every Mr. Olympia contest).

Part of the problem stems from greedy promoters of the sport who have used and discarded bodybuilders as they saw fit, merely to advertise their worthless products. Being rather more interested in their own pockets than in the welfare of the sport, these "popularizers" of bodybuilding have contributed to the negative image. For a while in the 1950s, publications appeared using photographs of professional bodybuilders that obviously catered to homosexuals. Until recently, bodybuilding was not a sport in which one could hope to make a decent living. Hence, bodybuilders were often forced to turn to such publications in order to support their families. However, according to Franco Columbu, author of *Coming on Strong* and a recent Mr. Olympia winner, bodybuilding ranks sixth among organized sports in terms of participating homosexuals—behind such "untainted" sports as football and tennis.

As far as the alleged shortening of lifespan is concerned, the facts show quite the opposite. Steve Reeves, Hercules of the screen and Mr. America of days long gone, is now well over 50 years of age and looking much the same as he did 20 years ago. The great John Grimek, Mr. America in the '40s, is now 70 years old and has remained essentially frozen in time for over 40 years. Bob Hoffman, a great weightlifting star of the '30s, is now nearly 80 years old and regularly lifts 250 pounds overhead with one hand at exhibitions.

Muscle stars, however, admit that there is some truth in the term musclebound. If bodybuilding is overdone and if sufficient stretching exercises are not included (e.g., chinning and toe touching), the muscles may contract and shorten their range of motion. Actually, this condition is very rare. The majority of people who seriously practice bodybuilding are sufficiently systematic in their approach to include a variety of exercises. In *Coming on Strong,* Columbu says: "I play table tennis and jump rope excellently to this day, which surprises people because of the myth that bodybuilders are muscle-bound, clumsy, and slow. No other sport on earth is anywhere near as rich in old wives' tales as mine."

As always, there is the perennial, "What happens when the body-builder stops working out? Doesn't all that muscle turn to fat?" These questions reflect a common misunderstanding: that ultimately the built-up muscle fiber will collapse into a sea of blubber. In fact this is a physiological impossibility. As Schwarzenegger put it recently, "To suggest that muscle cells could become fat cells is no more than biological alchemy." What can happen is that the muscle loses its "tonus", or resiliency. However, this will happen to anyone who ceases to exercise entirely and is not a condition peculiar to the bodybuilder. After a prolonged layoff, the bodybuilder's physique will return to pretty much its original condition.

The practical benefits of the sport are unquestionable. Jogging and tennis may be fun, but as health builders they cannot compare to body-building. The biggest argument in their favor is the so-called "cardiovascular effect" (exercising the heart and circulatory system by driving up the pulse rate for a brief period). Jogging will produce this (along with broken arches and bloody urine—as any marathoner will testify), but bodybuilding and weight training produce this effect every bit as much as jogging. Bodybuilding, like anything else, must be done intelligently. If the would-be athlete spends half his time in the gym yakking rather than training, he will reduce the cardiovascular effect and invite injury. The reason is simple: When allowed to rest too long, any muscle will "deflate" (blood leaves the tissue) and become cold. The muscle "forgets" it is supposed to be handling heavy weights and injury can result. This is the reason all trainers in any sport insist on the sacred warm-up period.

In a recent experiment conducted at Kent State University, some doubting researchers, among them Dr. Lawrence Golding, a well-known physiologist, subjected Mike Katz (Mr. Universe several times) to a battery of stress and fitness tests. They immersed Mike in tubs of water, took body measurements, did a stationary cardiovascular test, and monitored his pulse under a wide range of conditions. Every one of his scores was in the "high-good" to "superior" range.

An official medical committee report submitted to the International Federation of Bodybuilders concluded that muscle development actually improves agility, speed, flexibility, and endurance. The report further indicated that popular activities like jogging do absolutely nothing for the conditioning and develpment of the abdomen, back, shoulders, or arms. Many joggers suffer from lower-back pain—quite possibly the result of poor muscular development in that area. A *Time* magazine article (June '78) reported that 45 percent of women who jog extensively suffer from dysmenorrhea (painful menstruation). According to Dr. J. E. Schmidt of Charlestown, Indiana, in a recent *Boston Globe* article, "Running is one of the most hazardous forms of exercise for both men and women. Among the bodily structures most likely to be damaged by jogging are the sacroiliac joints, the joints of the spine, the veins of the legs, and the uterus and breasts. Nor are these the only casualties of jogging. Among others are the 'dropped' stomach, the loose spleen, the 'floating' kidney, and fallen arches. With a 'friend' like jogging, who needs enemies?"

Even before these studies, many colleges and universities used weight training to supplement training for other sports. Cornell University introduced systematic weightlifting as part of the varsity crew training program back in the early '60s. But the benefits of weight training on athletic performance go beyond colleges and universities. More recently, huge Lou Ferrigno (Mr. Universe and currently the star of the television series, "The Incredible Hulk") won his division of the "Superstars" competition defeating such "unmusclebound" athletes as Lynn Swan (Most Valuable Player in Superbowl X). Big "musclebound" Lou broke the record in the rowing competition, ran a fine 2:02.5 in the half-mile, and took second in the bowling and bicycling events. He placed seventh overall against the greatest athletes in the world.

Not to be overlooked are the mental advantages of the sport. Perhaps surprisingly, many bodybuilders find that the psychological benefits are as great or greater than the physical rewards. According to Arnold Schwarzenegger, in *The Education of a Bodybuilder*, "The process of bodybuilding does not, in my estimation, stop with the body. Seeing tremendous change can open new worlds for you. Plato wrote that man should strive for a balance between the mind and the body. Without a well-conditioned body, Plato felt that the mind would suffer . . . I have been able to create that balance within myself. I know that building the body makes the mind reach out."

Arnold Schwarzenegger

In the beginning there was Steve Reeves. But as the walls of Jericho came tumbling down, so did the career of the great star of the Hercules films of the '60s. Reeves' reign was far too abbreviated. With its passing, bodybuilding went back to the underground, where it lay dormant for over a decade. Yet the lull was temporary, for there was a mighty power gestating in far-off Austria that would soon burst upon the scene with a body surpassing Reeves' in both elegance and raw animal power. The era of Schwarzenegger had begun.

In a few short years, Arnold won the Mr. Olympia contest an unprecedented seven times, starred in several hit motion pictures and a television documentary, conducted seminars and exhibitions all over the world, and in general established himself as a one-man conglomerate in many phases of business and theatrical life. His rise to fame, prodigious strength, and captivating personality was so remarkable that his first and second books have become nationwide best-sellers. His first, *The Education of a Bodybuilder*, traces the odyssey of Arnold's life from his beginnings as the son of a small-town police chief in Graz, Austria, to the summit of world-class bodybuilding.

At 6 feet, 2 inches and 236 pounds, Schwarzenegger is not only the leader of world-class bodybuilding but the sport's most tireless promoter. Beginning bodybuilding at age 15, he became Mr. Universe just four years later. Nor is the Schwarzenegger physique pure "show." In addition to being a top-notch and formerly competitive swimmer, he

Arnold Schwarzenegger, bodybuilding superstar. (Caruso Photo)

has proven himself in powerlifting competition by hoisting 700 pounds in the deadlift—a weight four men of average strength could not budge from the floor.

Yet Arnold's prowess is not confined to the *biceps brachialis*. Like his colleague Franco Columbu (also a Mr. Olympia winner), his philosophy of life is avowedly Platonic. In his autobiography he declared that, like the great Greek philosopher, he believes that "The mind is incredible. Once you've gained mastery over it, channeling its powers *positively* for your purposes, you can do anything. I mean anything . . . 'I can't' should be permanently stricken from your vocabulary . . ." His career is overwhelming evidence that the latter has, in fact, long since taken place. Despite the fact that English is not his native tongue, Arnold continues to author articles for bodybuilding magazines both here and abroad. He handles the technical and financial aspects of his many businesses with ease. He is a staff writer and editor of *Muscle & Fitness* magazine and has been the subject of profiles and articles in publications ranging from *Cosmopolitan* to *The New York Times.*

Other dimensions of the great Austrian's philosophy of bodybuilding are both unexpected and controversial. Arnold convincingly demolishes the common myths surrounding bodybuilding by force of wit and sheer presence. One of the classic misconceptions is exemplified by the question, "What happens to the bodybuilder when he stops training?" Arnold's characteristically sardonic yet well-founded reply is that they simply stop gaining! In fact, that question is usually a thinly disguised expression of yet another common prejudice: "Won't all that much-vaunted muscle finally collapse into a mass of useless tissue, robbing the bodybuilder of his health, and dragging him inexorably to an early grave?" With a knowing smile, Arnold points out (probably for the zillionth time) that this is a physiological impossibility. As he and others have explained, to suppose muscle cells could become fat cells is to suggest nothing less than biological witchcraft. What *can* happen, Arnold readily admits, is that muscle cells and tissue can lose resiliency. However, he is quick to point out that this will happen to anyone who ceases exercise entirely. It is not peculiar to the bodybuilder.

Astonishingly, the great muscle star admits that there really *is* such a phenomenon as being musclebound. As he explains in *The Education of a Bodybuilder,* "The musclebound body is created by people who only lift weights and flex and contract their muscles, whose only thought is to get muscles. They never go beyond the flexing and contracting to the other movements the muscles need in order to stay supple." Specifically, if sufficient stretching movements such as chinning, running, and toe-touching are not done, the muscles contract and shorten their range of motion. But while this is theoretically possible, this condition occurs approximately as often as a debating loss for William F. Buckley, Jr.

Faddish nutritional programs and exotic zero-carbohydrate diets are elements of the bodybuilding scene for which Arnold has few supportive words. According to one popular approach to muscular supremacy, the aspiring Mr. America must first "bulk." That is, he must gain as much as 30 or 40 pounds of bodyweight beyond what he intends to ultimately carry. Some bodybuilders of the "golden age" of bodybuilding in the '50s, such as Bruce Randall (Mr. Universe, 1959), have gained over *200 pounds* more weight than they finally brought to the posing dais. The next step is to "cut"—to lose all but five to ten pounds of excess, leaving, in theory, nothing but pure, defined muscle. This is something Arnold would never advocate. The safest and surest way to a prize-winning physique, according to the great Austrian, is to either add weight slowly and scientifically if you are underweight, or to lose it the same way if you're overweight. To gain and lose according to fads or preconceptions rather than by listening to the rhythms of your own body is, according to Arnold, the surest way to ruin both physique and health.

The *right* way, in his philosophy, is to concentrate on the major food groups while simultaneously avoiding all refined foods. "Do yourself a favor," urges Schwarzenegger. "Start eating foods that will give you quality and vitality. Replace all refined sugar with honey. Avoid cakes, pies, candies, french fries and packaged snacks. Satisfy your sweet tooth with fresh fruit." Regarding the role of protein supplements, Arnold is less rigid and rightly so, for they generate great controversy among bodybuilders themselves. In *The Education of a Bodybuilder,* Arnold does recommend an extra protein drink for those in heavy training. On the other hand, his colleague Franco Columbu has declared, "Protein powders and liquid amino acid supplements are one of the biggest rip-offs in consumer history. They cause water retention and the protein source is usually collagen—a very poor quality protein." Perhaps under Franco's influence, Arnold does not emphasize the role of supplements in bodybuilding.

Brimming with pugnacious charm and immortal optimism, Arnold answers the hardest question of all: *why?* Why do "these guys" do this to themselves? What is the raison d'etre of bodybuilding? He answers that it is, for one thing, a form of art equal to or surpassing sculpture. In the movie version of *Pumping Iron,* he declared that sculptors have only to pack on a little clay, while bodybuilders do it the hard way; they work on the human body.

If nothing else convinces, Arnold cites the harmony of feeling that bodybuilding develops. Sacred to the bodybuilder is the "pump"—the infusion of hyperoxygenated blood deep into the muscle. It is a feeling that, like the aroma of coffee, is virtually impossible to describe. He once likened it to a sexual explosion, though that seems to be a mediocre analogy at best. Bodybuilders claim that it is one of the greatest sensations one can experience. Yet only a fortunate few have done so, despite its being no farther than the nearest gym. As Arnold said in his autobiography, "Plato wrote that man should strive for a balance between the mind and the body . . . Strength and confidence, plus a first-hand knowledge of the rewards of hard work and persistence, can help you attain a new and better life." Finally, as both Arnold and Franco believe, bodybuilding retards the merciless, pitiless onslaught of time in a way no other sport can. At 70 years of age, the immortals Sig Klein and John Grimek are living testimony to the faith.

Arnold retired temporarily (he won the 1980 Mr. Olympia contest) from competition in 1975, but he did not retire from bodybuilding. He continues to work out as vigorously as always and spends a good deal of time traveling and preaching the gospel of good health. He continues as a director of the International Federation of Bodybuilders' (IFBB) Professional Bodybuilding Committee, and he actively promotes the sport at home and abroad.

Perhaps the quintessence of Arnold's philosophy of life, art and bodybuilding is embodied in Columbu's observation that bodybuilding " . . . is a real art and science in itself. Many effective poses are based on ancient Greek sculpture and the later work of my countrymen during the Renaissance. We don't know for certain, but this is pretty fair evidence that our sport has really been around in some form for centuries."

Doris Barrilleaux

If Schwarzenegger and Columbu are the crown princes of male body-building, then Doris Barrilleaux merits the same status for the fairer sex. With an intellect like Curie and a physique like Aphrodite, she has single-handedly transformed the exciting new sport of women's body-building from something of an oddity to a respected and growing phenomenon. Despite her age—she is 49—she regularly competes with and defeats women half as old.

But it was not always so. Born in Houston in 1931, she was a thin, sickly child. Her bodybuilding potential really blossomed when she happened upon an issue of *Strength & Health* magazine, not long after the birth of her fourth child. Her early fascination with strength now rekindled, she soon mustered the courage to try her hand in a local men's bodybuilding gym. The women's era had begun.

In 1963 her first photos appeared in various bodybuilding publications. Though racked by misfortune and loss, including the loss of her sixth child and a subsequent hysterectomy, her iron will and enormous physical reserves continued to manifest themselves. She continued to work out, offer assistance to others, and travel widely. Soon she was expertly writing and reporting on the muscle world. After submitting photos to bodybuilding magazines, she began a friendship with the successful muscle photographer Dick Falcon. She began writing for the magazines and continues to do so today.

The friendship with Falcon was profoundly gratifying, but Doris became increasingly restless with the fact that bodybuilding was, as yet, a predominantly male enterprise. Though she had had success with her personal forays into the muscle universe, she wanted others to have the same opportunities. The next important catalyst came when she read of a woman's physique contest to be held in Ohio, in 1978. At 47, she once again displayed her exciting talent for the sport, placing a respectable third. The happiness, however, was tinged with displeasure: Though well-intentioned, the event was a study in chaos and disorganization. The alternative? There was nothing to do but to found her own organization. With the assistance of a local bodybuilder, Suzanne Kosa, they formed what is now known as the *Superior Physique Association* (SPA). With their superb organizational abilities, exacting criteria and rules were set up. The disheveled state of the art was now a thing of the past. *SPA News* was started to keep women informed of coming events, and SPA branches are now found throughout the United States.

Today, as president of the SPA, Region IV Chairman of the AAU, Executive Committeewoman of the IFBB, photographer and publisher, she travels widely and lectures to mass audiences on her first love. Recently she appeared on NBC's "Real People." Also, she has signed a contract with film producers for a documentary that will trace the evolution of women's bodybuilding throughout the world—sort of female *Pumping Iron.* Also, Doris has of late accepted an invitation to guide the women's section of the President's Council on Physical Fitness and Sport.

Of course, all is not smooth going for the pulchritudinous iron-pumpers. For one thing, there is the nagging question, "What is it?" A

Bodybuilder Doris Barrilleaux. (Doris. J. Barrilleaux)

response is not so easy to come by. Beauty contests offer tempting comparison. What is the difference? The skeptics argue that when male bodybuilding first crawled out from the dank, dark basements into the light of respectability, the analogy between the male version of the sport and a simple female beauty contest was obvious. Then, so the argument goes, if male bodybuilding was nothing more than a version of women's beauty contests, could female bodybuilding lay claim to any difference at all from the beauty pageant? The skeptical argument has had many adherents, the present author included. Indeed, before seeing and studying the philosophy and approach of the Superior Physique Association, my gut reaction was to dismiss it as a transparent attempt to mimic yet another aspect of the male world, solely for mimicry's sake.

My view now is rather different. As the Greek philosopher Heraclitus pointed out, all things are not as they seem. Initial impressions often stand in stark contrast to reality. This, the eternal problem of epistemology, lies at the heart of women's bodybuilding. Perhaps the simplest and most devastating difference separating women "iron-slingers" from the followers of the pageants is the fact that female bodybuilders have muscle and the beauty queens do not. Clearly, their muscular development cannot rival men's, though that is not the issue. The same philosophical points brought out in defense of male bodybuilding apply as well to the female version. The metaphysical essence of bodybuilding—be it male or female—is to bring the musculature and health to the state of development nature intended. The beauty pageants are an absurd conglomerate of mashed-potato thighs, asinine attempts of self-analysis, and the most profound ineptness masquerading as talent. Women's bodybuilding, by contrast, relies not on "hype" and plastic smiles, but on pure physical development—the universal aesthetic quality readily appreciated by all.

Nor is this philosophical difference lacking in practical distinctions. Rather than trying to execute some intuitive, unformed, and uninteresting conception of beauty that greatgrandma might love, women's bodybuilding has brought the concept of female beauty into the space age. Barrilleaux, Laura Combes, and Charonne Carpentier strive for an essentially Aristotelian notion of form, grace, and mathematical proportion. It is not surprising, therefore, that the contestants rely heavily on weight-training—the finest and only true road to physical perfection. Not only is the female bodybuilding contestant far more muscular, better proportioned, and better developed than the old-style beauty contestant, she is also healthier. (It would be most interesting to put Doris Barrilleaux in head-to-head competition for fitness and beauty with the "Miss America" winners and runners-up of about 22 years ago as they look *today!*)

As with any emerging sport and art, there will be diehard skeptics. Pockets of resistance remain here and there, even with men's bodybuilding, though the awesome Arnold Schwarzenegger demolished all the stupid myths years ago. The inimitable David Susskind, for example, recently gave the valiant ladies a hard time when Doris and three other lady bodybuilders appeared on his show. As Ms. Barrilleaux explains, "To our dismay, we were immediately aware of the fact that Mr. Susskind had two primary interests and both were located on Lynn's (Johnson) upper anatomy . . . In fact, whenever April, Lynn or I attempted to interject anything into the program we had a difficult time turning the conversation from sex."

Still, adherents of women's bodybuilding ought not despair. Evolution takes time. The public reception of the art will improve. As Doris herself explained in a recent issue of *SPA News*, "A year ago no one believed there was a place for women's bodybuilding. The few who did believe it worked very hard to prove to the world there was a place for women's physique competition. Now that the women have been so well accepted worldwide and have their own group well organized and are expanding everywhere, others are now jumping on the bandwagon . . ."

Amen and good riddance to the days of Betty Grableism, airbrushed cellulite, and tinselly "beauty" contestants who would have difficulty doing Zottman curls with a Bert Parks balloon.

2

Olympic Weightlifting: The High and the Mighty

Montreal, 1976. He might, at first glance, appear to be just another over-weight businessman, yet he is engaged in no businessman's task. At 6 feet, 2 inches, and 345 pounds of awesome strength and surprising agility, he is the strongest man alive and the very essence of the contemporary sport of Olympic weightlifting. He is about to try to raise, to arms' length overhead, a barbell loaded to a monstrous 562¼ pounds—a weight no human being in history has ever lifted. To the arrogant Russian it might as well be a broomstick. Unlike his competitors, he spends little time "psyching" himself, thereby telegraphing to the world that the weight is far, far from his limit. Within seconds he squats to grasp the bar, rises, and rams the immense load of iron effortlessly overhead. His competitors pale. Alexeev: He has no equal. He is the Babe at Yankee Stadium, Bobby Fisher at Reykjavik, Graham Hill at Indianapolis.

Who are Alexeev and his kind? What is this sport that drives men to demand more of themselves than one lifetime can provide? They train, diet, and sleep with one and only one goal—to lift as much iron as possible over their heads. As top-rate American lifter Mark Cameron put it, "Listen, if you told a weightlifter he could lift five more pounds by eating Brillo pads, there wouldn't be a clean pot anywhere in the world." A competitive lifter will spend up to five hours per day doing countless numbers of deep-knee bends with up to 700 pounds on his shoulders, bench presses with up to 500 pounds, and overhead lifts with up to 400 pounds. He will practice form and technique for hours in front of a mirror using nothing heavier than a yardstick, and will consume as much as 400 grams of protein a day (a normal man needs about 70 grams).

A favorite of many competitors is a high-protein "milk shake" consisting of massive spoonfuls of protein powder, several raw eggs, huge dollops of honey, molasses, whey, wheat germ, peanut butter, a few

David Rigert, Soviet weightlifter. (Novosti Press Agency)

bananas, and some ice cream thrown in for good measure. These are all mixed in a blender and consumed in gargantuan gulps.

Sleep is extremely important: Most pre-competition lifters will get 10 to 12 hours per night. This allows for the regeneration of vitality and muscle tissue broken down during a workout.

Immediately prior to a competition a lifter may do nothing for days but eat and sleep. Abstinence from exertion, including sex and routine exercise, is common. Some lifters become reclusive in this period, hoarding carbohydrates that will burn like coal during a meet. In the lighter weight classes, where staying within body-weight limits is crucial, the pre-competition regimen will be determined by the ease or difficulty with which a lifter expects to make the weight. For the superheavyweights, however, there is no such concern. They will break up their rest periods with enormous eating binges, packing on as much bulk and protein, and as many carbohydrates, as they can. There are legendary stories about meals eaten on the morning of a major competition. For example, a few years ago in London a frisky Alexeev is said to have eaten a whole side of beef and an entire layer cake *after* his normal breakfast. Then there was the moving performance of American Bruce Wilhelm before the 1978 Senior National Championships. After his breakfast he settled his stomach with 40 pancakes and a dozen eggs before going back to bed.

Competitive Olympic lifting, as it is practiced today, consists of the "snatch" lift and the "clean and jerk." The snatch is by far the more difficult of the two; here the lifter must bring a barbell from the floor to arms' length overhead in a *single,* quick movement. In the clean and jerk, the lifter first brings the barbell to his shoulders (the "clean"), pauses a few moments, and then tries to jerk the barbell to arms' length overhead.

Often, the training for these lifts reaches proportions matched only by Alexeev's diet. Typically, a weightlifter concentrates on developing three things: power, speed, and technique. Training is always well planned and scientific. The average lifter will spend three days a week in lifting during the non-competitive part of the year. A typical first-day routine might consist of practicing various phases of the snatch lift (such as snatching the bar overhead, without trying to rise) and various phases of the clean and jerk (such as working on the clean alone without trying to jerk the weight). On the second day, a lifter will often work on those phases of lifting not practiced on the first day. Thus he might practice jerking a weight off a rack (eliminating the need to clean the weight) and doing deep-knee bends with the bar held overhead, as if he had just completed a snatch. This way he builds both technique and power. On the third day he will try to "put it all together," practicing the lifts in their entirety. Many lifters practice pure power movements, such as full squats and presses. Usually they work on their lifting form during all three days, using very light weights. "Light" is a relative term, of course. Alexeev and Rakhmanov practice repetition cleans and jerks with about 290 pounds. For the confident Russians this is about equal to a normal man working out with a heavy crowbar.

Historically, weightlifting was not always as precisely defined nor taken as seriously as it is today. Indeed, the roots of modern lifting go back to antiquity, spreading throughout many cultures and many lands. The lifting of heavy stones was a popular pastime in ancient Greece, with the names of athletes traditionally inscribed on the stones they succeeded in hoisting. The Greeks had their heroic strongmen, just as we have the great Alexeev now. So revered was the great Milo of Greece that the philosopher Aristotle refers to him often. In his *Nichomachean Ethics,* for example, Aristotle mentions Milo in illustrating the doctrine of the "Golden Mean," according to which every person, if he is to achieve happiness, must find the intermediate point or "mean" in any choice in life. Lest one think that the great philosopher was concerned solely with nobler things, he takes great pains to urge that eating, among other activities, must be done in moderation, citing the eating habits of Milo as that which would be extreme for the normal man, though quite in order for Milo.

Obviously, Aristotle recognized the outlines of modern training methods because the great Milo, partially due to his great appetite, was virtually invincible in hand-to-hand combat, having won six Olympian, nine Nemean, and ten Isthmian games in his career. And, of course, there was his legendary carrying of the bull on his shoulders every day until the animal was full-grown. (Even then the Greeks knew about the virtues of progressive resistance!) So great was Milo's prowess, in fact, that wrestling was finally eliminated from the old Olympics in Milo's seventh appearance because, as a philosopher wisely noted, "neither god nor man durst stand against him!"

There were, as one might surmise, other athletes of note in ancient Greece. Polydamas may well have been the largest athlete of ancient times, dwarfing even some of the giants Plato discusses in his myths. According to some guesses, he was nearly seven feet tall and weighed more than 300 pounds. There was Theagenes, who lived around the time of Plato; his place in the halls of the immortals of strength is assured by the records indicating that, at nine years of age, he walked off with a statue of an Olympic athlete that may have weighed in excess of a quarter ton!

Nor were the Greeks the only culture boasting superheroes. In ancient Persia, classic tales tell the story of Rustam, the mythological folk-hero. In one myth, Rustam is said to have fended off legions of demons using strength of limb alone. In *Tales of the Hero,* it is stated that "Rustam, using his great strength, pushed his attack to the utmost. Time and time again the great war club crashed against the head and vitals of the monsters, gouging out great masses of hide and flesh. Even the floor of the cavern shook under the impact of the blows. So murderous was the assault that not even this greatest of demons could long withstand it, and soon he sank to the floor battered and helpless."

Then, of course, there is Sumo, that noble and least bloody of all contact sports, practiced by the Japanese for centuries. The classic story of Sumo is inexorably rooted in great strength. As the legends have it, a deity, unpretentiously referred to as the "Deity-Brave-August-Name-Firm," met another humbly named weightlifter called the "Brave-Awful-Possessing-Male-Deity." Appropriately enough, the contest opened with a weightlifting gambit. Allegedly, the former deity came "bearing on the tips of his fingers a thousand-draught rock" (a rock it would take 1000 men to lift). Unfortunately, the show of power didn't do much, and the rocklifter ended up pleading for his life.

Nor did the shows of strength stop with the mythical conquests of Japan. Later it happened again; a show of lifting power saved, or helped to save, the empire. As the story goes, Shiranui, the eleventh and one of the fiercest *yokozuna* (the highest title in Sumo wrestling) hoisted a 140-pound sack of rice onto his head, put three more on his shoulders, and carried the load around for hours to apprise Commodore Perry's soldiers of the fate that would befall them upon invading Japan.

In the Middle Ages, when St. Augustine was rewriting Plato and St. Thomas Aquinas was plagiarizing Aristotle, supermen of strength continued to arise. Edward II, king of England in the 14th century, is said to have used a sword so heavy that no single man could wield it. (Some writers have claimed that this sword is still to be seen in the Tower of London, but I was unable to verify that on a recent visit.) Likewise, Sir

William Wallace, a legendary Scot of the 14th century, used a sword that, allegedly, he was the only man in the world capable of wielding.

As the centuries passed, filled with war and international bickering, strongmen continued to flourish. William Joyce of Kent, a strongman who lived in the 17th century, regularly demonstrated his feats before King William III at Kensington Palace. Among the most amazing of these was a staggering four-part feat of muscle. First he lifted a solid lead object weighing around 225 pounds; thereupon a horse was given the chance to try to pull him from his spot (unsuccessfully); thirdly he broke the rope previously attached to the horse; and finally he smashed the posts previously straddling the ropes (i.e., the rope had been tied between two posts when he broke it).

In the 18th century, innumerable feats of lifting strength were recorded. William Ball of England is said to have carried a metal object weighing half a ton across the length of the foundry room, while a colleague in muscle, William Carr, carried a half-ton cannon for half a mile!

Despite the long heritage of strength and the fascination with lifting seen in so many lands, modern weightlifting has to be counted as a relatively recent phenomenon. It became popular around the end of the 19th century and was noted for the presence of many British and European supermen. Why this should be so is not hard to understand. The late 19th and early 20th centuries was an era of great optimism, and optimism always breeds strength. Change, moreover, was everywhere. The telephone entered the scene in 1885. The automobile, railway systems, refrigeration, generally improving economic conditions, increased mobility, and a positive anticipation of the future contributed to growing public confidence.

As the musclemen headed for our shores, the public flocked to see the wonders of nature in action. Circuses, music halls, cafes all featured strongmen and they all contributed to an American fascination with strength. Add to all of this the fact that the first modern Olympic Games occurred in 1896, and you have the conditions that would eventually give birth to an Alexeev. Yet, despite the presence of weightlifting in the first modern Olympiad and despite the victory of the great Elliot of England and Jensen of Denmark, modern weightlifting remained rooted in the public consciousness as a vaudeville fixture full of tawdry tricks and stunts.

In the '20s, the legendary Sig Klein, for example, stupefied audiences by holding a girl in one outstretched arm over a bed of razor-sharp knives. The great French-Canadian, Louis Cyr (about whom more will be said later), is alleged to have hoisted over 4300 pounds in the back or "platform" lift in 1895. On other occasions, Cyr is said to have raised 16 men, or more than 3500 pounds, in a back lift. As with so many such claims, it is difficult to separate fact from fiction. Still, given Cyr's enormous power, such feats are far from out of the question. In another, somewhat more reliable report, Cyr bested strongman Dave Michaud, who had been claiming to be Canada's strongest man for years. Cyr easily won, hoisting a boulder that weighed nearly a quarter ton—a boulder none of the other strongmen could move at all.

Eventually, weightlifting moved out of the dance halls and into the serious lifting auditoriums. Unofficially, according to weightlifting sage Bob Hoffman, serious competition emerged in the United States primarily because of the efforts of Father Bill Curtis about a century ago.

Interestingly, Father Curtis, good Thomist that he was, emphasized the health-building aspects of weightlifting, as modern strongmen such as Franco Columbu do today. (Probably modern enthusiasts would call that weight "training" rather than weightlifting, in keeping with modern distinctions.) Whatever one chooses to call it, Father Curtis got things rolling. He was instrumental in the formation of the Amateur Athletic Union (AAU) in the latter part of the 19th century, as well as the promotion and systemization of weightlifting.

Another important step occurred when Henry Steinborn introduced the so-called 'quick" lifts, the one-hand snatch, the two-hand snatch, and the clean and jerk, the latter two being the only historical lifts officially sanctioned today. Developments in organization and in the lifts themselves followed rapidly. In 1928, Sig Klein approached the AAU and persuaded them to include weightlifting in the USA under their umbrella, thereby relieving the crumbling *American Continental Weightlifting Association* of the enormous dissent, financial difficulties, and political infighting that had been eroding the vigor of the organization. Still, all was not well in the glory-days of the century. Despite the noble tradition of vaudevillian strongmen, organizational progress, and an enormous will to survive, the spirit of American weightlifting could not dredge up enough talent to enter a team in the first AAU competitions.

In 1932, on the other hand, enough new blood had arrived to give the USA team strength enough to enter the Los Angeles Olympiad. Dick Bachtell and youngster Tony Terlazzo entered the featherweight class, Wally Zagursky and Arnold Sundberg the lightweight division, and massive Howard Turbifill the heavyweight division. Though gutsy, the line-up was still less than worldbeating. Only Terlazzo managed to defeat any of the European lifters, placing a respectable third. No one could suspect that, by the end of World War II, the Americans would be supreme in the world—for a while, at least.

Organization and promotion continued and, by 1936, the United States team had matured. Tony Terlazzo again was the standout, improving on his own best performances to garner a gold medal, while setting world records in both the press and three-lift total (which then included the press, as well as the snatch and the clean and jerk; the press was done away with in the 1970s). It was in that year as well that the immortal John Grimek burst into prominence by setting a national record in the press and beginning his road to bodybuilding immortality as well. The next year Terlazzo, in his turn, took the spotlight from Grimek by accomplishing the unprecedented feat of winning a world championship in another weight class (the 1936 record being in the featherweight division, while the 1937 mark was in the lightweight). Possibly the most exciting year in American weightlifting was 1938, when the young and relatively inexperienced Steve Stanko defeated another soon-to-be immortal, John Davis, to capture the national junior championships.

The premier moment in American lifting history occurred that same year when Stanko, going right on to the Seniors, joined the ranks of the mythical heroes of old by raising an unbelievable pre-steroid 1000 pounds in the Olympic lifts. One cannot help wondering how many of today's lifters could, without anabolic drugs, duplicate such a feat. Following such triumphs, however, the American forces could not maintain the pace. Though there were great talents, such as Harold Sakata, later to gain fame as "Oddjob" in the James Bond movies, other nations were exerting their lifting might. Though the United States team captured four gold and two silver medals in the 1952 Helsinki Olympics, the promise of the post-war years remained unfulfilled. Yet there were signs in the firmament: prophecies abounded of a great unseen force about to burst forth. Soon the great Tommy Kono emerged, eventually capturing his share of fame by establishing 25 world marks in four different weight classes. And, indeed, names such as Schemansky, Ike Berger, and Kono, in and of themselves, would have sufficed to ensure the place of the United States in lifting history.

But the talent and achievements of such great men were all but erased from the public psyche with the sudden eruption of one of the very greatest forces ever seen in the history of Olympic weightlifting, Paul Anderson. The leviathan from Toccoa, Georgia, easily established himself as the most powerful being in recorded history, to that time. The unimaginably powerful "Dixie Derrick" obliterated all past accomplishments in superheavyweight weightlifting before retiring to run a boy's home in Georgia, to which he continues to devote full attention today.

There are many who remain profoundly convinced that Anderson could, if he so desired, pulverize the greatest lifters today, despite his nearly 50 years of age. This is a heavy claim, given the presence of the Soviet goliath, Vasili Alexeev. It is a romantic notion to be sure—a battle between Anderson and Alexeev. Such an encounter would be apocalyptic in every sense of the word. We think of the great Paul Morphy returning to do battle with Bobby Fischer over the chessboard, Jack Dempsey stalking Muhammad Ali at Madison Square Garden, the Babe versus Catfish Hunter at Yankee Stadium. To argue over the winner of such hypothetical encounters is, perhaps, only slightly less whimsical than speculating on the outcome of the battle of Armageddon.

Anderson's strength was almost mystical. In his early years he raised 6,000 pounds in the back lift—the greatest weight ever lifted by any human being. He has hoisted automobiles as if they were sacks of flour, pushed freight cars as one might push a shopping basket, twisted railway spikes in the manner of a child amusing himself with taffy—all seemed to be child's play for the great Anderson. During a famous showing of the Ed Sullivan Show, Anderson hoisted nearly 5,000 pounds of awkward, live human weight in the back lift. That lift seeming to be his specialty, the immortal Paul Anderson decided to break the record of the French-Canadian lifter and strongman, Louis Cyr. Until Anderson's era, Cyr had held the record in the back lift (a lift done by crouching under a platform loaded with weight and lifting it with the hips and back) of an incredible 4000 pounds. With the assistance of a supportive father, Paul began the assault on Cyr's record. First they constructed a table which in itself weighed 1800 pounds. When the apparatus was finished and loaded, Anderson lifted it with no difficulty, exceeding Cyr's record by more than a ton, and lifting far more than the weight of a conventional automobile.

Doubtlessly, the finest moment of Anderson's career came in the Melbourne Olympics of 1956, with an accomplishment of sheer Christian faith and determination that might make the most convinced atheist have second thoughts. The competitions had not been going well for

Paul Anderson, the leviathan from Toccoa, Georgia. (Paul Anderson Youth Home)

him; for days he had suffered with fever. Thirty pounds down in body-weight, swallowing aspirins like candy, he developed an inner ear infection and could barely walk because of balance disturbances. Both the press and the snatch had not gone well. By the time of the clean and jerk phase of the competition, Anderson was in the unenviable position of needing a new Olympic record to compensate for the other two lifts. He needed 414½ pounds to take first place. It was a potentially disastrous situation. Just a few weeks before, he had lifted nearly 500 pounds in practice and would have toyed with a mere 414½ pounds. The first of his allotted three attempts was a dismal failure, and the second was worse. Only Anderson's own words can capture the pathos of the situation, as the desperately sick man demonstrated the strength of will and faith that moved him.

"In 180 seconds I felt the presence of God. I walked up a long, dark corridor and felt as if God was reminding me of everything He had ever done for me . . . The arena was silent . . . When I pulled the weight to my chest in the cleaning motion, I knew immediately that it was futile. I couldn't put it overhead. Now I was desperate. In a split second I found that I could be sincere with God . . . I bent my knees slightly for the momentum to push the weight. The bar had already stayed on my chest from four to five seconds, an unusually long time for that amount of weight . . . I drove the bar overhead and it stayed."

Yet, by 1960, the career of the fabled "Dixie Derrick" had ended as abruptly as it had begun, when the Amateur Athletic Union stripped him of his amateur status for accepting a few dollars to appear on television. The loss of Anderson cost the American forces a power which they have so far failed to regain. In the Rome Olympiad of that year, Chuck Vinci grabbed the gold for the US in the 123-pound class. This was to be the last gold medal a US citizen would win in the Olympics up to this printing. By 1964 the eastern-bloc nations and the Soviet Union were rising to world domination. Only Ike Berger and Schemansky could manage a silver and bronze, respectively, against the Soviet supermen. Results were scarcely better in the 1968 Olympiad, when only Joe Dube could grab a third place in the heavyweight division. The '72 Olympics are best forgotten, so far as United States Weightlifting is concerned.

What is the state of Olympic lifting today? It would not be inacurrate to say that the US is strong, though optimism has to be kept in perspective, in light of recent injuries and retirements of top American competitors. Until very recently we had Mark Cameron—a world-class lifter who managed a fifth place in the Montreal Olympics. But it is beginning to look like he may be retiring. In the same weight class (242 pound), however, Guy Carlton is extremely strong and should fill the void left by Cameron. Among superheavies, old-timer Joe Dube is returning powerfully to the platform and, of course, there is the matter of superheavy Tom Stock. He recently suffered a massive leg injury and there is a real question as to whether he will ever lift again.

Still, the Soviets have their problems as well. Ordinarily there would be no question whatsoever of Alexeev's dominance, except for the fact that he bombed completely in the Moscow Olympics. Since then, much has been made of the failure with many predicting the end of Alexeev and the beginning of a new reign by Sultan Rakhmanov, who took the gold in Moscow. It's quite possible that such talk is premature, however.

It should be noted that at 38, Alexeev is still far from "old". America's Norb Schemansky was competing successfully into his forties. It should also be noted that many top powerlifters are well into their fourth and even fifth decades. Legendary John Grimek, for example, still squats with over 400 pounds, though he is over 71 years old! In my view, a kind of psychology has developed among Olympic-style lifters; many feel, without any evidence, that "40" is some sort of death-figure for the olympic lifter. I believe the lifting longevity of many historical strongmen and the examples from powerlifting refute that idea.

All this is not, of course, to disparage Rakhmanov's fine performance in Moscow. He is clearly superior to Alexeev in the snatch. However, his clean and jerk in Moscow was, according to some eyewitnesses, the very limit of his abilities in that lift and very far from Alexeev's best. And Alexeev was, by all accounts, badly out of shape when he showed up in Moscow. So if he can avoid the losing psychology caused by the number "40," he could be on the winner's podium for many years. In fact, I'm still confident that the Soviet giant will eventually hoist an unbelivable 600 pounds in the clean and jerk. And there is no reason to doubt that future lifters will raise records considerably higher. How far can a human being go? My view is that one day, someone, somewhere, will lift 700 pounds overhead. This, of course, is only speculation.

What the future of world weightlifting holds cannot be predicted with certainty. Chances of the US regaining world domination are problematic because of the resources of the state-funded East European athletic programs. Shotputter Beyer and weightlifter Bonk are examples. By contrast, American athletes, weightlifters especially, must either pay their own way or depend upon the assistance of the patron saint of American weightlifting, Bob Hoffman and others. Indeed it is surprising that we have had lifters of the caliber of Stock and Cameron in recent years. If the United States can return as a superpower in weightlifting, it will not be because of the very great efforts of Hoffman, the prodigious strength of Stock, or even the excellent coverage and publicity afforded by the networks. Ironically, the final credit for this country's resurgent interest and competitiveness must go to the power, presence, and prestige of the great Soviet colossus known the world over simply as Alexeev.

Vasili Alexeev

"Yech-h-h. This caviar is not so good as at home. But I really don't like to eat. I get sick when I think of food."

Thus spake Vasili Alexeev, superheavyweight weightlifting champion and the strongest man of all time. Still he recently managed to devour the following: two chocolate cakes, two legs of lamb, several platefuls of peas, endless kegs of beer, tons of caviar, rice, fruits, roast beef and dressing, and salads of every conceivable variety. The great man was visiting the famed Carvery Room of the Tower Hotel in London en route to Montreal when he devoured this record-shattering fare.

Recently I had the opportunity to interview the waiter and several waitresses of the great dining hall and they talked freely about the immense difficulties involved in keeping the gregarious giant satisfied. According to the chief waiter, "He nearly bankrupted us. Every night he consumed a whole side of beef, several cakes, potfuls of potatoes, beans, puddings, and washed it all down with barrels of beer. I believe I shall never witness the likes of such a being again!"

We can only conclude that Alexeev's disclaimer about his eating habits is a feat comparable only to the awesomely heavy weights he has been hoisting overhead for over ten years now. With his incredible power, and nimbleness to match, he is truly, as the Soviets themselves once said of Paul Anderson, a "wonder of nature." There have been many in the history of the "Iron Game" who have applied themselves as diligently, but few who could even begin to match the accomplishments of the Soviet Hercules. Despite the recent claims of some powerlifters, "Uncle Vasili" (as he is fondly referred to by his fans) has to be considered the strongest man in recorded history. For powerlifting is still a young sport with years ahead of it before the full limits of human potential will be approached, not to mention the fact that no powerlifter has yet met the great Alexeev in a test of strength. Additionally, we know

The Soviet superhuman, Vasili Alexeev. (Novosti Press Agency)

that the great Vasili as well as other Olympic lifters have reached or bested the finest efforts of the best powerlifters. (Paul Anderson has done full squats for repetitions with over 1200 pounds, a weight no powerlifter has ever handled. And even Anderson never approached Alexeev in Olympic lifting!) Of course the shadow of Anderson remains, as a contemporary battle between Alexeev and Anderson at full strength and with modern techniques would be a most difficult match.

Six feet tall and weighing 345 pounds, Vasili Alexeev, despite recent failures, still considers himself unbeatable. So do many of his competitors. His record of total domination began in 1970, when he became the first man to clean and jerk over 500 pounds in competition. Since then he has held every world record at one time or another, with the total now exceeding eighty.

Before the Montreal Olympics the future was not so bright. For one thing, there was the great Bulgarian Khristo Plachkov, who snatches more than most heavyweights jerk. Then there was Gerd Bonk of East Germany. Indeed, just weeks before the Montreal Games, Bonk had stolen the clean-and-jerk record away from Alexeev in the European Championships with a jerk of 252.2 kilos (557 pounds). To make matters worse, the upstart Bulgarian, Plachkov, only 23 years old and far from his peak, had stolen the record for the total in the two lifts with a combined snatch plus clean and jerk of 442.5 kilos (976 pounds). Oddly, the great Alexeev was bothered not a jot. When his coach, Plukfelder, was asked of the whereabouts of Vasili just days before the competitions, he answered nonchalantly, "At the lake, fishing."

For the fact was that Uncle Vasili had reserves of power within his huge frame that no one could dream of. When asked of the challenge of Gerd Bonk of the German Democratic Republic, Alexeev's consummate self-confidence thundered forth in his now famous reply, "Bonk could not even beat me if I were 50 years old." And indeed Alexeev may be blessed with the gift of prophecy as well as overwhelming strength. In the days just prior to the great Olympic Games, the opposition was crumbling before the charisma of Alexeev like so much ash in the wind. Plachkov had knee problems and was not lifting well in training. Only Bonk had the temerity to remain in the competition. Unfortunately, though Bonk was a fabulous clean and jerk man, his snatch was not quite on a par with the best superheavies (though it is vastly improved today). Playing it safe, Bonk took only 375 pounds in the snatch, while Alexeev opened with 386, finally winning the snatch competition with 408 pounds. (Interestingly, Plachkov at his best can approach 450 in the snatch!) Then came the clean and jerk, Alexeev's and Bonk's forté. Once again Alexeev waited indifferently until the other superheavies had finished. Starting with an easy 507, Bonk actually bettered that on his last try. However, with a total of 415 kilos (915 pounds) to Bonk's 405 kilos (893 pounds), the gold medal was safely in Soviet hands.

At this point, Alexeev's real reason for coming to Montreal emerged. The scoreboard announced that Alexeev would attempt an amazing 255 kilograms, or 562 pounds—by far the heaviest weight any human being had ever raised to arms' length over his head. Immediately, the crowd erupted in frenzied cheering and wild applause. Alexeev strode onto the stage to take his rightful place as ruler of the Olympic universe. Even the great Bonk himself, for all his might, could only stare wide-eyed at the imperturbable Russian. For an eternity Alexeev leaned against the wall, eyes closed as if summoning all the Forces of Darkness to assist him. Presently he approached the bar, emitted a thundering "Yarg-h-h-h," flipped the massive load of iron to his shoulders and then immediately slammed it overhead, the titanic tummy shaking and quivering. His comment on the effort? "Maybe now my wife will show me more respect."

The hopelessness of competing against Vasili was summed up by American lifter Bruce Wilhelm: "Any time you compete against Alexeev, you know you cannot win." Nor were these idle words: faced with the reality of the continuing presence of the Russian, Wilhelm has since retired.

The future of Alexeev is unclear because of that thigh muscle tear he suffered in a recent world championship tournament. Many authorities are still convinced that the Soviet giant will someday hoist a mind-boggling 600 pounds in the clean and jerk. Should he accomplish this, it would surely be the most incredible athletic feat in history—comparable to running a three-minute mile. Until his injury problems are resolved, however, question marks hover over the indomitable spirit and future of the great Russian and, by definition, the future of Olympic weightlifting, of which Alexeev is the Platonic embodiment. For it must

Alexeev grappling over 500 pounds. (York Barbell Company)

be noted that at the time of the Moscow Olympics, Vasili was only 38 years of age, a time when most weightlifters are just coming into their prime.

Still, Uncle Vasili has other things to do while waiting for a worthy competitor to challenge him. (Sultan Rakhmanov temporarily bested an out-of-shape Alexeev in the '80 Olympics but in top form Alexeev is much stronger.) As indicated earlier, one way he passes the time before great competitions is by fishing the great lakes and streams of the world. Also, he is nominally a mining engineer in the Donbas region of the USSR. Man of mystery that he is, Uncle Vasili allows his fans to wonder about far more than just how much weight he can put over his head. Of equal wonderment is the amount of weight he has been able to put around his stomach. Indeed, many esoteric theories about the role of the mountainous midriff have circulated in recent years—elegant theories about "distribution dynamics," "centers of gravity," "balance," etc. But Alexeev explained that, "I eat, because it is necessary for my sport. But I nearly throw up every time I think of eating."

Of course, maintaining that kind of hypertrophied flesh is not possible by only eating. An Alexeev workout is said to be the eighth wonder of the world. While his wife, the aptly-named Olympiada, looks on, the big man can be seen going through endless repetitions of heavy squats, power cleans, presses and snatches. He will warm up with a mere 300 pounds on the bar in the squat—easily completing 25 or 30 repetitions "just to get the blood moving." Eventually he will work up to as much as 800 or 900 pounds in this movement, with little or no apparent increase in strain. Like Charon guarding the secrets of Hades, the Soviet Superman never gives any hint of what his limits, if any, might be. Even his teammates are kept in the dark. As the equally remarkable middleheavyweight lifter David Rigert observed recently, "I myself am perplexed about Alexeev's training. He seems to spend most of his time fooling around with weights which, though they would surely crush the life from an ordinary man, are no more than toys for his amusement."

Though certainly not verified, it is rumoured that Alexeev has done squats with something near 2000 pounds on his mighty frame. We can only begin to grasp the enormity of such a feat by noting that this is the equivalent of going from a standing position to a full deep-knee bend and rising again with a small automobile slung over the shoulders! "Nobody can beat me," he says. "I can go on for years." Though the recent Olympiad has, unfortunately, temporarily negated that claim, the boast is modest for Alexeev!

Uncle Vasili is equally colorful, titanic, and unpredictable offstage as on. Between rounds, Alexeev can be seen striding imperiously among the masses of admirers in whatever capital he is visiting. Looking very much like just another ponderous construction worker, his celebrated golf hat, black t-shirt, and the 60-plus-inch waist dropping proudly over a massive belt buckle, the Soviet superman gazes disdainfully about him. He is supremely oblivious to the meager specimens of humanity around him like Godzilla wondering which part of the city to wade through next.

One such uneasy host to the great man was New York City, that arrogantly proud and hostile gotham, the city which, to its inhabitants, is the sum and substance of existence. But that day New York was put in its place. He started with that yippy shoe salesman on 46th Street (you know, the guy who ignores you for an hour and then asks if your feet are clean?). Well, the big Russian humbled him by mightily hoisting him above his head with one hand, so that he might try on a pair of shoes.

But even Superman can crack a joke. Alexeev relaxed his guard a bit when a couple of female models wished to climb onto his twenty-inch biceps. "Don't send any of these pictures to my home," the great voice intoned. (Similar photos were dispatched to Olympiada in 1970; luckily, Alexeev survived.) There is further evidence of his sense of humour. His interpreter, Yuri Radzievsky, recently furnished one anecdote to *The New York Times*. "Someone asked Vasili if he did much running. 'Only if someone is chasing me,' he said." Alexeev then wryly pointed out that turtles never run, and they live to be 300 years old (a good point for those tortured souls among us afflicted with the jogging mania).

"How do you like New York?" a gaper inquired of Uncle Vasili. Displaying his usual sound judgment and appropriate reverence for the fabled city, the great man answered simply: "The same as I did ten years ago." (Unlike many die-hard New Yorkers, Vasili apparently feels that it takes more to justify one's existence than residing within the confines of the Hudson and East rivers.)

What's next? For the present, Alexeev's injuries and mysteriously poor performances cast a dark shadow. We can only wait. If eventually he does recover his earlier form, the tables will turn. The weightlifting world will be faced with the prospect of defeating a being whose power defies human explanation—a power that is almost biblical. If that's a fair analogy, then in order to unlock his secrets we shall perhaps need a modern-day Delilah whose task it will be to "Entice him, and see wherein his great strength lies, and by what means we may overpower him, that we may bind him to subdue him."

But first she'll have to get past Olympiada

3

Powerlifting: Human Derricks in Action

"If you're going to lift it, *lift* it!" Sad comment. Still, in the minds of untutored onlookers in the throne room of the gods of powerlifting, that is how the sport strikes them. In the minds of many, to lift a weight is synonymous with lifting it over your head. Hence, the deadlift is often viewed as a negative accomplishment—the awesome force required to bring the defiant weight merely to waist-height just emphasizes how very, very far the competitor is from being able to hoist such a poundage overhead.

In a way, the sport is, well, scary. It is a glimpse into the future. We watch the great Alexeev hoisting 562 pounds overhead with ease, and cherish the hope that he, or some other human volcano, will finally crack the 600-pound barrier. Yet to watch the equally great Don Reinhoudt struggling mightily to bring 900 pounds barely to his knees is seen, paradoxically, as a symbol of defeat, though the lift be successful. If one as awesome as Reinhoudt—a being so huge and powerful that one greater is nearly unthinkable—can barely raise such a weight to the waist, has he, even in the moment of victory, unwillingly shown us the very limits of human power? For how, in the light of so monumental a struggle, can we expect a single human being to hoist such a load of iron *overhead*?

No less an authority than Clyde Emrich, Olympic lifting great and strength coach of the Chicago Bears, has alleged that powerlifting, unlike Olympic lifting, requires much less athletic process, technique, speed, agility, and, in fact, courage. Why courage? Anyone who has ever tried Olympic lifting knows the terror a weight can cause, when you are faced with the task of raising it to arms' length overhead. But with the deadlift, according to Emrich, the fear is minimal: If the weight resists one's mightiest efforts, well, just drop it a few inches to the floor with nary a stubbed toe.

33

Vince Anello deadlifting 821 pounds. (Powerlifting USA)

Still, I exaggerate. For powerlifting is a wonderful sport and, undoubtedly, the purest expression of raw human muscle in existence. Unlike Olympic lifting, the sport requires virtually no technique, speed, grace, or form. Some will argue that it is an inferior sport for that, but this is an assumption without foundation. If raw power is part of the essence of man (or woman), then the sport crystallizes that element and is as fully deserving of the designation "sport" as any on this earth.

As practiced today, powerlifting competition consists of three events: bench press, deadlift, and the squat. Historically, the sport goes back to antiquity, though it has existed as an organized activity with its own governing bodies only since the early '60s. Certainly, the practice of lifting heavy objects goes as far back into recorded history as we can discern. In anticipation of the modern-day deadlift, for example, the great Greek wrestler Milo (5th century BC) was reputed to have lifted rocks, statues, and other massive weights clear of the ground. The Greeks generally practiced the "dead" lifting of heavy stones, and stones have been discovered inscribed with the names of strongmen who were said to have moved them. The practice of lifting heavy stones was carried into the Middle Ages as well. In Munich, the *Apothekerhof* castle guards a huge boulder—estimated at 400 pounds—bearing an inscription suggestive of ancient Greek times. The inscription (which actually is on the adjoining wall, rather than on the rock itself) indicates that Duke Christopher of Bavaria gave resounding proof of his manhood by lifting the giant rock (a practice similar to the raising of the culturally important "manhood" stones in contemporary Scotland).

Similar practices go on today in the Basque regions of France, Spain, and parts of the American west, where giant stones are "dead" lifted for high prizes. Additionally, there is an annual stone-lifting competition held in Munich, where a boulder exceeding 500 pounds in weight is attempted by the hardiest German youths (and an occasional oldster).

But such bizarre lifts bear only distant resemblance to modern-day deadlifting. And if we look for historical roots for the other powerlifts, the bench press and squat, there is precious little. A major reason for this is the fact that these modern lifts became popular only with the advent of evenly balanced barbells and dumbbells, developed toward the end of the 19th century (as was the case with the modern Olympic lifts as well). Even then, however, as well as in the so-called "Golden Age" of strongmen in the '20s and '30s, there was relatively little to suggest modern-day powerlifting. What we find instead is a strange array of the bizarre and exotic—one-hand overhead lifts in contortionist postures, barrel-lifting, lifting barbells with the teeth, etc.

In the '40s and '50s, however, an age of scientifically designed benches and interchangeable barbell plates (in contrast with the fixed-weight monstrosities of vaudeville) began, and the practice of powerlifts gained wide acceptance. One of the earliest to exhibit a phenomenal rise in popularity was the bench press, so much so that the '50s produced some of the largest, and most overdeveloped pectoral muscles in the history of the Iron Game. This lift may well be the single most popular and most widely practiced movement in the world of strength. It is incomparable in the satisfaction, invigoration, and ease of performance it offers, producing some of the fastest strength gains possible. It is not uncommon for a total beginner to go from a low of 100 pounds to well over 200 pounds in this exercise in a few months.

Though the equipment used in gyms varies somewhat, the basics are simple. The lifter lies on a bench and lifts a weight from the chest to locked arms' length overhead. In competition this is precisely defined. According to AAU rules set in the '60s, the bench must be 12 inches wide, 18 inches high, and about four feet long. The contestant usually has the weight handed to him, lowers it to the chest, and holds it for two seconds. He then attempts to push it to arms' length again.

The deadlift, the most basic test of overall strength, is simplicity itself. The lifter merely grasps the bar and rises in a continuous motion so that he finishes in an upright position, shoulders slightly back and knees locked.

The "squat" is really a modern term for the old-fashioned deep-knee bend, with added weight on the shoulders. Like the deadlift, this also is a basic movement in lifting: no glamour, technique, or aesthetics to mar its brutally elemental nature. The lifter positions himself under the bar-bell, which is usually supported on a rack (bar no lower than 1.2 inches below shoulders), lifts the weight clear of the rack, and descends slowly into the squat. God willing, he rises again, the hostile poundage con-quered. Besides its presence in competitive powerlifting, it is renowned as a fundamental developer of body power, often causing the muscles of the legs, hips and buttocks to reach apocalyptic proportions. Virtually every weight thrower, bodybuilder, and strength athlete of any worth has practiced the movement at one time in his or her career.

I indicated that only the deadlift can be said to have any sort of historical root; still, with some stretching of the concept, it could be argued that the squat maintains ties with the past as well. It has been realized since the time of Plato's *Republic*, for example, that athletes have performed the classic deep-knee bend—not so much as a power builder but as a creator of endurance and general fitness. Many exotic strength athletes around the world practice the movement as well. The mudpit wrestlers of India, the Basques of the American west, and the giant *Sumatori* of Japan all make like Atlas reincarnate.

Nonetheless the squat, as practiced with the added resistance, re-mains a very recent phenomena as sporting events go. The remarkable Henry "Milo" Steinborn (named, apparently, after the great Greek wres-tler of old) brought the lift to the United States. After Steinborn's time, however, there was really no practitioner of the squat worthy of men-tion until history collided with the fabled Paul Anderson of Toccoa, Georgia. Anderson has time and time again proven himself one of the most powerful—possibly *the* most powerful—human ever to walk the earth. He is on record as having performed 1160 pounds in that lift. Unfortunately, no one seems to know for sure what the great Alexeev of Russia is capable of in that move. Occasional unverified reports have it that Alexeev has done over 1100. If Anderson has done 1160, then 1100 for Alexeev is eminently believable, given that Alexeev has a physique similar to Anderson's in terms of leg structure and hip strength. Additionally, Alexeev has hoisted well over 500 pounds overhead, far more than Anderson ever did.

Of course, it doesn't follow that either Alexeev or Anderson is the greatest squatter in history, since lifters in the lighter classes have done proportionately more. David Rigert of the Soviet Union is reported to have done three consecutive full squats with nearly 700 pounds, at a bodyweight of just under 198. If we extrapolate from this and assume (which seems reasonable) that Rigert could do one full squat with 800 pounds with a little emphasis on this lift, this might well surpass Ander-son's performance (as would Mike Bridges' awesome 826 lb. world record, set recently in the same weight class). There is, again, no official verification of Rigert's squatting ability. But, very likely, the squatting capabilities of many lifters in the smaller classes, if seriously tested, would exceed Anderson's pound-for-pound.

The bench press is synonymous with Pat Casey. Pat, who set a record of 617 pounds while others were struggling to bench 350, must be regarded as *the* immortal of history. Though others have since done more, it was Casey who opened the door. Also, Doug Hepburn of Vancouver must be noted. Hepburn performed a bench press of 580 pounds in 1954 (which, if he were using the narrower hand grip required today, would most likely have been no more than 500). Also worthy of mention in the recent history of the bench press is powerful Mel Hennessy of Minnesota, who benched 560 while tipping the scales at barely over 215. And the great Marvin Eder amicably tossed up over 500 pounds back in 1953, weighing less than 200 pounds himself.

Interestingly, the smaller Olympic-style lifters have done some phenomenal things with this lift. The diminutive but powerful Chuck Vinci, Olympic lifting champion in the bantamweight class in 1956 and 1960, has rammed up over 300 pounds while weighing in at barely over 120. Not to be left behind, the bodybuilders have powered up absolutely enormous poundages without serious specialization in the lift. Harold Poole, noted physique star of the '60s, hoisted an outstanding 425 in 1966, while weighing just under 163 pounds. Franco Columbu, the popular and genial star of *Pumping Iron* and holder of all the major titles in bodybuilding, cannot be overlooked. At a bodyweight of barely over 180, he is an inhumanly strong powerlifter, having shoved 485 pounds to arms' length in fine style.

Moving into the specialists in the lift, it is extremely difficult today to single out anyone, the movement being as widely practiced as it is. People like Doug Young, Vince Anello, Don Reinhoudt, Terry Todd, Ron Collins, and Larry Pacifico, to name just a handful, have all performed outstandingly. New powerlifters are appearing daily. At the *Olympia Gymn* in Woburn, Massachusetts, for example, a titan named Ernie Hackett is smashing superheavyweight records, despite the fact that he tips the scale at a "mere" 275 pounds. Though a bit behind the very best, Hackett is improving fast. It must be noted, however, that powerlifting is a new sport. Records come and go like Soviet gymnasts. Much more is in the offing.

It is precisely this fact that plays havoc with the best efforts and eternal desire to compare powerlifters with their Olympic brethren. Who is strongest? Though such comparisons are approximately as easy as asking whether Capablanca would have defeated Fischer in chess, it is asked so often that it must be dealt with. A great deal of nonsense has been written about both powerlifters and Olympic lifters, and few hard facts are available. For one, there have been relatively few head-to-head encounters between representatives of the two groups. On CBS's "World's Strongest Man" competition, Bruce Wilhelm, an American Olympic lifter, did tussle with a couple of powerlifters, including big Don Reinhoudt. Wilhelm did indeed win that competition, defeating big Don in a final tug-of-war. Now if one were to take that meeting as representative, it would appear to indicate that Olympic lifters are stronger than powerlifters, the reason being that Wilhelm was, by world-class standards, a relatively minor figure, while Reinhoudt was probably the best powerlifter in the world. However, some degree of "psyching" may have occurred, as Wilhelm's line of annoying chatter is as big, figuratively speaking, as are his biceps. Most strength authorities I've talked with feel that Reinhoudt is actually far stronger than Wilhelm. In any case, this was, as indicated, a single clash, an insignificant moment in history. And though powerlifters and Olympians have met other times in similarly silly "contests," there has been, to date, no controlled, scientific battle between the best of the two groups. It would be most interesting to see the great Alexeev and America's Reinhoudt in head-to-head battle. Could the Soviet superman match the American's 885-pound deadlift? Personally, I believe he could, though I would not like to place a *great* deal of money on it.

Terry Todd—Powerlifter

In the course of planning this book, I collided full force with one of the biggest moral problems of my life (well, almost): who to profile in powerlifting. Affable Don Reinhoudt, with his deadlift of two-and-one-half times his bodyweight? Marvelously muscled Marvin Phillips, with his phenomenal 777-plus pounds world-record squat? Or Terry Todd, founder and midwife of the sport of powerlifting and holder of the first official 700-pound squat (performed as a junior)? After considerable soul-searching, I decided in favor of Terry Todd. The choice was logical. Without the ground-breaking efforts of Dr. Todd, there would have been no records for his descendants to break.

At the first AAU-sanctioned power meet, held in York, Pennsylvania, in 1964, Todd wore the hats of both organizer and competitor. Though steeped in his doctoral studies at the Unversity of Texas, he managed to win the superheavyweight division with ease. Equally meritorious, undoubtedly, was Todd's breaking of Bob Peoples' world-record deadlift of 725 pounds—a mark that had stood the test of time and tendons for over 15 years. (Some writers have tried to disparage Todd's accomplishments, urging that Paul Anderson could easily have demolished Peoples' record years ago, not doing so only because they were friends. Obviously this is possible; quite obviously it is also possible that many others might have broken it. The fact is that Todd *did* break it.)

Todd's career as a powerlifter evolved out of a fine career as an Olmypic lifter. Weighing around 240 pounds, his first experience with that sport was as an undergraduate at Texas. Eventually he worked his way up to the National Intercollegiate Championship and became the junior national champion in 1965.

Eventually, however, Todd grew to such proportions that he became literally "musclebound" as far as the Olympic lifts were concerned. (Yes, Virginia, the condition exists. Franco Columbu told me it can arise with really enormous size.) His arms, particularly, got so large that he began to have trouble holding and cleaning heavy weights. And with thighs approaching 37 inches in girth, it was evident that the man was made for the powerlifts. Not surprisingly, he went on to officially total 1900 pounds in the sport.

Todd retired in 1967 with a record of total domination equalled only by Alexeev's accomplishments in Olympic lifting. Nevertheless, he continues as a univeristy professor and writer, having authored many articles for various muscle magazines. With his powerful wife Jan, Terry Todd works both as a farmer and a professor at Dalhousie University. Eschewing modern technology as far as possible, the Todds grow their own vegetables, raise their own pork, and heat with a wood-burning stove. It is easy to get the impression that the Todds are waging total war with modern industrialization and our "plastic" environment. Not so: no Rosseauian turnback to a primordial state of nature for the Todds. "We are moderately self-sustaining," says Terry. "No; make that *somewhat* self-sustaining. That would be more accurate."

Jan Todd

There is a pair of stones in Scotland. No, there is *the* pair of stones in Scotland. These mountainous boulders are the rocks upon which are built all great Scottish legends and feats of superhuman power. For centuries they have lain by the river Dee, defying all who would attempt to raise them. For any man to lift them was considered the ultimate test of *machismo*—mankind at full throttle—but for a *woman* to dare the feat? One might as well ask Satan himself to apply to the Boy Scouts. Yet this amazing woman, wife and matriarch of the Todd clan, did precisely that. The awesome stones that defied the strongest men in Scotland for millenia have been conquered by the fairer sex. As a *Sports Illustrated* journalist recounted her dramatic feat, "I saw her face flush as she lowered her hips and began to pull, and I shouted along with everyone else as the smaller stone came up quickly, much higher than before, and I shouted again as she leaned back and at long last the larger stone swung clear. It came off the ground neither very far nor for very long, but by God it came . . ." Too much was too much. This feat, along with superhuman Bill Kazmaier's powering the 268-pound "Inver" stone (another of the Scottish stones of strength) over his head, caused noted Scottish Games authority David Webster to remark in goor-natured desperation, "Och, but I've seen it all now. Yesterday the Dinnie stones conquered by a woman, and now the Inver Stone handled as if it were a pebble."

Yet the hoisting of the famed Dinnie stones is but one milestone in the illustrious career of Jovian Jan Todd. A college graduate, Jan is amazonian in proportions, beautiful, and staggeringly intelligent. She is, along with husband Terry, a vigorous and articulate defender of her sport. Like Terry, her name is synonomous with powerlifting. Without

Powerlifter Jan Todd, who has bested the famous Dinnie Stones. (Kathy Tuite)

her efforts there would be no woman's powerlifting. Always naturally powerful, Jan used to entertain people by bending bottle caps between her fingers before she even approached her current level of might.

Still, Jan was and is interested in many other things. She met husband Terry when they were both at Mercer University, a Baptist liberal-arts college in Georgia. She edited the college paper, reached the highest academic levels, and generally kept the college running smoothly. Near the end of Jan's senior year, she and Terry began to work out together, slowly at first, since Jan was at the time more interested in general conditioning than setting records. Eventually they were married, with Jan yet to be bitten by the "iron bug."

Soon an incident occurred that forever changed her attitude toward the weights—and the course of her life. While she and Terry were working out at the Texas Athletic Club, Jan witnessed a wisp of a woman— bodyweight no more than 115 pounds—gradually dragging up an unbelievable 225 pounds in the deadlift. Following conversations with the woman about training, diet, competition, etc., Jan decided that if so diminutive a female could compete successfully in men's contests, she could do likewise. Also impressive was the fact that the smaller girl, train like a maniac though she did, had succeeded in piling on enormous strength without adding the highly visible male-type muscularity much feared by women. Of course, it is well known that women do not, by and large, develop the quality of muscle that men develop, primarily because the muscle-producing hormone testosterone is present only in minute quantities in the female system. (There are women in the sport who do tend to show male-type muscularity; however, it's also known that many such women are taking muscle-building anabolic drugs. This, of course, tends to turn female lifting into a weak imitation of male bodybuilding and weightlifting. Jan Todd, on the other hand, has never taken such drugs.) A few years ago, this natural fear of "muscles" did hold many women back from competitive powerlifting. Once aware that no such thing would happen to her, Jan threw herself into the sport with abandon.

The first of her many staggering accomplishments was the erasure of Mlle. Jane de Vesley's deadlifting record from the record books. According to the records, Vesley had pulled a whopping 392 pounds back in 1926. Having lifted 225 virtually her first time out, Jan felt she would smash the record with ease. Immediately, with the assistance of her equal half, Terry, she concentrated on the most basic power movements—bench press, squat, and deadlift. Surprisingly, she did not meet the natural male hostility one might have expected, as she moved out of the cellar and into the big-time gyms of Georgia. To the contrary, perhaps recognizing her great natural abilities and extraordinary intellect, the other lifters accepted her immediately, offering the same encouragement they would have given to any other trainee. Her progress was phenomenal. By the fall of 1974 she was within range of 400 pounds in the deadlift, and was beginning to think seriously about the Vesley record. Finally, exactly one year and four months after first learning of that mark, she pulled 394½ pounds and forever erased Vesley's name from the books.

But that was only one of the many victories. Soon the social activist in Jan began to surface as she pondered the state of powerlifting. In its short history, the sport had been dominated entirely by men. When women did enter, they had—like the diminutive female who first interested Jan in the sport—to lift against men. Nor were they always warmly welcomed. Horror stories abounded; stories about women having to strip in front of judges (to make certain there were no illegal support bandages) or having to wear jockstraps because "the rules required it"—travesty after travesty. Jan immediately made up her mind that she would have none of it and, with the cooperation of Joe Zarella (himself a veteran powerlifter and organizer) introduced the first national powerlifting competition for the fairer sex.

However, before her triumphal day on that, the first of the national contests for women, obstacles (perhaps planted by some recalcitrant male sorcerer) lurked behind every Nautilus machine. At Christmas, 1975, she slipped on some icy steps at a friend's farm, thereby incurring a severely sprained ankle. An injury like that is extremely serious to a powerlifter, since the ankles are the reference point for the two heaviest lifts in competition, the squat and deadlift. A sprained shoulder might interfere with bench pressing, the lightest of the power lifts, but can be worked around in the heavier movements. Still she kept at it and by springtime she decided to try for a world-record attempt in a powerlifting meet in Nova Scotia, coming through brilliantly with 412.5 pounds in the deadlift.

Slowly the ankle healed, and for the first time in her career Jan began thinking about the other lifts. The first women's event she and Zarella had helped create was looming over the horizon. Systematically, she and Terry planned her training. She had hoped to become the first woman to total 1000 pounds in the powerlifts, a feat that would have rivaled the late, great Steve Stanko's historic totalling of over 1000 pounds in the three Olympic lifts. But the omnipresent enemy of all weightlifters, Dame Gravity, stood close to her side that fateful day.

After succeeding beautifully with a 385-pound squat and a 170-pound bench press, the great lady dragged 445 pounds of groaning, sagging iron to within centimeters of success. It was my impression that had she taken the 445 immediately, instead of draining herself with an earlier 415 (which she conquered handily), the greater weight would have sailed up. It is difficult to realize how fast the powerlifts can rob one of vital energy, especially in the tense atmosphere of competition. After lifting a total of more than half a ton to that point, she was nearing exhaustion. Of course, winning the meet rather than setting records is the first order of business in competition, so her strategy was unimpeachable. She did win that day, stealing the hearts of millions.

A word is in order at this point: It has been my experience that the uninitiated can be unimpressed by, say, a 440-pound deadlift. "She's only getting it off the floor, after all," they scream. Worth noting, however, is that "only" lifting such a tonnage is quite beyond the powers of over 99 percent of non-weight-trained men, and a fair percentage of weight-trained men. I've seen many a burly 250-pound plus male fail to even budge such a weight. In a recent contest held at Boston State College in which this author participated, the winning deadlift in the middleweight class was less than 500 pounds. A deadlift of 500 pounds or more will win many regional and even state meets in men's competition in the heaviest divisions, if matched with a good bench and squat.

Ann Turbyne, world-class shotputter and lifter. (Jim Grass)

But gravity has not been all gravy for big Jan. Her supremacy has been challenged by a female behemoth named Ann Turbyne. A world-class shotputter, with a torso like Schwarzenegger and legs like Alexeev (well, not quite), the aptly-named Turbyne has lately been winning tourneys left and right. In a recent competition in Nashua, New Hampshire, this Madonna of Muscle powered up a mindbending 236-plus pounds in the bench press, 424 pounds in the squat, and 457½ pounds in the deadlift.

Whether Jan can turn back this superhuman woman, despite her own recent 506-pound world-record squat, remains the question today. Drugs, of course, might solve the problem, giving her the added boost to turn back the challenge of Ann Turbyne. To date, she continues to resist the temptation. As Terry Todd explained recently, "Yes, some women might use it, but not Jan." He went on to explain how rhesus monkeys at the Yerkes Primate center in Atlanta, when given the muscle-building anabolics at early stages of pregnancy, experienced profound physical and mental changes. "I am not ready for that yet," he adds. "I would like Jan just the way she is."

4

Sumo Wrestling: Muscle, Fat, and Philosophy

According to Japanese legend, the Sun Goddess sealed herself in a cave because she was irritated with her brother. As a result, the skies darkened and the world waited as furious attempts were made to release her. Among the more imaginative of these were a dance and a crowing cock. When she peeped out to investigate, the Herculean god Taukarao clapped his hands, stamped his feet, and pulled open the ponderous stone door with sheer force of limb.

Thus did one of the great traditions of Sumo wrestling—the stamping of feet and clapping of hands—emerge from an ancient legend of strength. There are many such, marking Sumo as one of the paradigmatic sports of strength in the world today. At an early stage in the evolution of the sport, tradition has it that there lived in the village of Taima a soldier and fighter of awesome might (Kuyehaya), renowned, according to the *Nihongi* (Japanese chronicles), for his boast that "you may search the four quarters, but where is there one to compare with me in strength?" (A somewhat premature boast as it turned out; the emperor sent for a warrior who promptly dispatched the fighter with ease.) The *Kojiki* (record of ancient matters) tells of the mythological conquest of Japan where a deity, bearing a rock so heavy 1000 men were required to lift it, asked, "Who is it that has come to our land, and thus secretly talks? if that be so, I should like to have a trial of strength." From this a wrestling match ensued between the rock-carrying *"Deity-Brave-August-Name-Firm"* and a challenger unpretentiously named the *"Brave-Awful-Possessing-Male-Deity!"* The rock-lifter promptly demonstrated that he needed to lift some more rocks and, according to the historian Hikoyama, "The divine race took over the land and thus succeeded in effecting an important national and social unity of the race." Hikoyama further comments that "the spirit of the wrestler, as well as his generous attitude of mind to pardon the vanquished, who surrendered himself and his all, is worthy of special attention as constituting the spiritual element in the way of the wrestler." With legends such as this, it is no wonder we find the wrestler in modern Japan regarded as much a philosopher as an athlete.

Still, Sumo evolved only slowly from a chaotic, no-holds-barred activity to the precisely defined and highly ritualized events seen today. Approaching its modern form in roughly the middle of the 18th century, the form and rules have changed comparatively little in today's arena. And in the 20th century, Sumo became widely appreciated as a spectator sport for the masses (as opposed to the religious function it had had for a relatively limited audience of royalty and clergy). With the advent of radio and television, there is hardly a person in all of Japan who does not know the latest promotion to *Yokazuna* (the highest rank in Sumo). Obtaining anything but standing room in famed Kokugikan Hall before a top-flight match is an accomplishment worthy of the Brave-Awful-Possessing-Male-Deity himself!

Matches take place in a ring of raised dirt, about 15 inches above ground level and nearly 18 feet square. This *dohyo* is filled with 28 earth-filled bags made of rice stalks. The bags are embedded in a circle on the top of the mound, forming a ring 15 feet in diameter. Prior to each

Sumo is a highly formalized but dramatic sport. A match is decided by one wrestler throwing his opponent or pushing him out of the ring, 15 feet in diameter. A bout seldom lasts more than two or three minutes. (Consulate General of Japan, N.Y.)

match, handfuls of salt are thrown into the dohyo by the contestants, following the ancient Japanese custom of purging all impurities with salt. Most wrestlers merely toss a small quantity of salt carelessly. A few of the more ambitious toss huge handfuls over the spectators or even themselves. In fact, the myriad ways of tossing salt have become trademarks of some combatants. (Awesome "Tubby" Wakachichibu was called the "Big Salt" in the late '60s because of his many theatrics in the salt ritual.) Still, the sprinkling of the salt, however it is bandied about, does have a serious mythic and religious function. In olden days it was believed to be anathema to evil spirits, preventing defilement. More practically, many wrestlers claim it keeps the mat surface hard and has medicinal effects on the cuts that frequently occur in Sumo jousting.

Interestingly, there are ceremonial events as early as the entry into the ring. Though only the top-ranked wrestlers take part in these colorful displays, they are an integral part of the whole Sumo scene. The wrestlers march down a flower-strewn path in single file, the highest ranks in back. They are then announced, one by one, as the top-ranked giants assume their rightful place in the center of the group, with the others facing them. Though this often functions as little more than a roll call (as well as a Sumo fashion show!), it is said to have evolved from the days when Sumatori fought as teams, offering prayers before battle. (Some have compared the ritual to Western oath-taking at athletic events; it is worth noting, however, that the Sumo rituals are always rooted in theological principles, while the Western ceremonies are inevitably secular.)

The fight itself may take only a minute, though the elaborate ritual preceding the bout may last as long as four minutes. Along with stamping their feet, the Sumatori will clasp hands, bow, open the palms, point to the audience, shuffle, do deep-knee bends in a prayerful attitude, and rub the body with paper—often with the effect of unnerving the opponent as much as appeasing the gods! An interpretation of the omnipresent foot-stamping, for example, is that it expresses the intention of the combatant to crush out his opponent, as well as the somewhat less terrifying "evil spirits."

No discussion of the ceremonial aspects of Sumo could be complete without mention of the elaborate symbolism and philosophical implications of the ring itself. It is claimed that the circular ring inside the square represents infinity, or chaos. And this, in turn, implies the philosophy of Ch'i—the two principles of Yin and Yang symbolized by the entrances on both sides of the platform. (My conversations with Sumatori indicate that such interpretations are often trotted out more for the benefit of impressionable tourists.)

As indicated, the actual Sumo bout is relatively short. It is decided when one of the wrestlers is pushed, carried, or dropped out of the ring, or brought to the ground. Yet what goes into and surrounds those fleeting moments is rich with psychological warfare, muscle, and tradition (as well as considerable fat). The psychology of a top-ranked Sumo bout is a wonder to behold, often eclipsing the savagery of the bout itself. Not too many years ago it was common for two gargantuan warriors to join heads, reminiscent of the deadly horn-locking of warring bull antelopes, and stare at one another for over an hour, the fans thundering approval while the warriors worked themselves to a frenzy. Nowadays, a time limit is imposed on this psychological battling, though it remains an essential ingredient of Sumo. During even those shortened moments, every manner of preparation imaginable may go on. A few Sumatori will, only at that last dramatic moment, calculate the strategy they intend to use. A few less confident combatants will indulge in prayerful meditation (especially, presumably, when the opponent outweighs them by a couple of hundred pounds, which is occasionally the case).

The bout itself, despite its brevity, often makes use of an enormously large number of throws. The professionals have a repertoire of over 200 techniques. (This figure is disputed by some authorities; the amateur Sumo federation of Japan lists less than 100. Apparently it is something of a philosophical question as to when a slight variation on a throw constitutes another kind of throw entirely.) Of these, by far the most difficult in the total Sumo repertoire is the *Tsuridashi*. It involves lifting the opponent by the sash he wears around his waist and carrying him from the ring—a feat only the very strongest wrestlers can employ, given the enormous size of these giants. Hence, the explanation of the ponderous, quivering bellies of the greatest wrestlers! It is not so much to maintain balance as is sometimes thought, but to combat this terribly deadly technique.

Another deadly, though very risky, move is the dreaded *Utchari*. Often this technique is an absolute last resort for a warrior on the brink. If he sees that he is gradually being overpowered, he will actually accelerate his own backward fall out of the cherished *dohyo*. But as he topples, seemingly to his doom, he hoists his would-be vanquisher and twists him quickly to one side. The result, if successful, is a study in Elizabethan drama: The wrestler within micrometers of defeat hits the floor, but with his "conquerer" actually touching down an instant before, thereby losing instead. In fact, this move can almost be considered a trademark of the great Taiho, 47th grand champion of Sumo and holder of most of the records in the sport. On the few occasions where he was in danger of being pushed out of the ring, his uncanny timing, technique, and superhuman strength would culminate in a successful use of the *Utchari*. Time after time the great man brought low his opponent that fraction of an instant before Taiho himself toppled.

The myriad maneuvers in Sumo are grouped into pushing, thrusting, and grappling moves. Inevitably, a bout between all but the most unequally-matched Sumatori requires all of these techniques. In the initial seconds of a bout, the Sumo contest may often more resemble a boxing match than wrestling. After charging each other with the ferocity of renegade bull elephants, the opponents pummel and punch each other (always, importantly, with open hands) with blurring speed. The point of such lightning thrusts is not to "knock out" the opposition, as is the case with the less civilized "sport" of boxing, but merely to unbalance him, so that the other maneuvers (such as *Tsuridashi*) may be employed effectively.

If the Japanese gargantua movies were inspired by the sport of Sumo, the Japanese art of *jiu-jitsu* gained vigor because of the famed *Uwatenage*. This throw, which is reminiscent of the throws of jiu-jitsu, characteristically involves pulling the opponent's sash by gripping him firmly from the outside and tossing him down. Because of leverage and balance considerations, this move requires more technique than pure power (though that element is always present in Sumo). Perhaps even more reminiscent of some jiu-jitsu moves is a throw known as *Kotenage*. Here the Sumatori reaches behind the opponent's left (or right) arm, hooking forward below the elbow and pivoting to carry out the toss. No grasping of the sash is involved in this case.

Unlike many Western sporting events, the referee plays no small role in these mini-psychodramas. Steeped in tradition, the *gyoji* is as respected in classic Sumo as the soldiers themselves. Generally, the referee determines the winner. In borderline cases, however, a final decision is made by four inspectors, each placed strategically on one of the four sides of the ring. Indeed, in modern times, even the four judges are not all-mighty. The "boob tube" has pierced the cloistered world of Sumo. In the biggest matches, TV cameras are placed strategically around the ring. Surprisingly, their presence is applauded by the tradition-conscious Japanese. According to chief judge and Sumo watcher Takasaga, "Television is very helpful to us. I recommend it to US football."

Garbed in their typically flowered Sumo robes and ornate coloration, many referees come to be known and loved. Through successive accomplishments, officiating first at bouts between lower-ranked wrestlers, they gradually work their way up the Sumo pyramid. So important is the role of the referee that, like the wrestlers, he can be promoted or demoted by a series of brilliant calls or silly errors. A referee's decisions, if overturned too often, can cause him to be dropped back to officiate among the younger aspirants. To symbolize increasing wisdom as well as progress through the ranks, the garb of the referee changes as he moves upward to the apocalyptic battles between yokozuna. At the very height of his powers, the gyoji wears the quintessential Sumo color of purple, as well as a wooden sword (to facilitate, according to some accounts, the committing of *hari-kari* in the event of a drastically wrong decision).

An oddity—explicable only in the wonderful and exotic world of Sumo—is that the thousands of referees in Japan are allowed to have only two different names. Every gyoji, regardless of rank, is named either *Kimura* or *Shikimori*. Shikimori, meaning "guardian of Kimura," must perforce be the underling to the latter. This fact derives from a legendary referee of several centuries back, who was considered so much the virtuoso of his trade that it was decreed that all chief judges must henceforth be named Kimura. Only when the Kimura dies or retires can the minion be promoted to Kimura.

Today all professionals belong to the *Japan Sumo Association*. The latter holds six major tournaments each year in Tokyo as well as in many of the other larger cities. Indeed, the Emperor himself is an enthusiast of Sumo.

Naturally, Sumo is not restricted to spectators and the professionals. There is an extremely large following among school children and others with no professional status or aspirations. A number of amateur tournaments are sponsored every year by high schools, colleges, industries, and other non-professional organizations. Though many participate in Sumo purely for the enjoyment of the sport, the more promising youngsters are always being sought by Sumo scouts, much in the style of American baseball recruiting. The reward for catching the eye of a trained scout is very often an invitation to become a professional.

Surprisingly, there has been increased interest in Sumo wrestling in other countries. Two of the greatest names in the sport have been only partially Japanese. The great Taiho had a Russian father, and the contemporary colossus Takamiyama (Jesse Kuhaulua) is an American citizen of Polynesian ancestry. Born in Hawaii in 1944, the great Takamiyama is an ancestral cousin to another great Hawaiian muscleman, Eddie Corney. But it is to the great Sumo star and his lovely birthplace that we turn next.

Takamiyama

There is a pervasive feeling in the West that Japan, despite the enormous gains in industrial virtuosity since the war, remains an essentially feudal, tradition-bound culture, rigidly closed to outsiders. Nothing could be further from the truth. Even as this is being written, two non-Japanese idols dominate the pages of the papers, as well as the airwaves of the great islands.

Sadahara Oh, Japan's answer to Mickey Mantle and Louis Tiant, is the famed first baseman of the Yomiuri Giants baseball club and surely one of the greats of the game. Though of Chinese origin, normally anathema for the Japanese, Oh has fought his way to the very top echelons with his legendary feats of long-ball hitting.

Still, it might easily be thought that Oh is, well, playing baseball—some sort of cultural impurity in the otherwise stable and eternal Japanese order of Truth. Yet the theory collapses, precisely because the quintessentially Japanese cultural phenomenon of Sumo has, for the last decade, been dominated by the mighty charisma, force, and presence of a non-Japanese superman named Takamiyama. Now well into his thirties, the Hawaiian-born giant (real name, Jesse Kuhaulua) is among the heaviest, oldest, and most loved of all Sumatori. In contrast to Sadahara Oh, who suffered the pangs of racism early in his career, the great Takamiyama has not, happily, encountered such problems. Not to suggest, of course, that the transition from the easygoing pace of the reclusive and serene Isle of Maui to the regimentation of the Sumo world was trivial either. But so cognizant of his own overwhelming talent was Takamiyama that he swept aside such barriers as easily as he does the mightiest of opponents.

A natural powerhouse, he learned the basics of Sumo from a Japanese coach while still in Hawaii, though at the time his interests extended to other sports as well. He played football while in high school and could put the shot over 50 feet. Predictably, the gargantuan size and strength of the young Takamiyama caught the eye of leaders of a visiting Sumo tour and young Jesse was offered a chance in the heady, fast-paced, high-stakes world of big-time Sumo. He never looked back. The charisma of Sumo may have forever welded itself to young Jesse's consciousness because of the presence of Takasago—one of the all-time greats—in the original party that first spotted Jesse.

Chances of success were small. Nearly everything—language, customs, environment—were alien to Jesse in the Sumo universe. He may as well have landed on Jupiter. Even his great physical size, once thought an asset, seemed to work against him in the early days. Sumo experts by the score predicted the early collapse of his career, contending that only Japanese possessed the needed physical structure and fighting qualities. Mere size and strength, they argued, would simply not be enough. And early experience bore this out, for a while. Jesse was culturally shocked at having to rise before dawn in the Sumo stables (training camps for aspiring wrestlers) and by having to cater to the whims of older, more established Sumatori. Even the food conspired to defeat him. So different was the fare of the grimy, no-nonsense stables, that Mrs. Takasago had to prepare special Western dishes until Jesse could acclimate himself to the local "haute cuisine."

To add injury to dietary insult, the great one lost most of his voice during a practice session when a training partner jabbed a tad too vigorously at his voice box. Warrior that he is, Jesse chose to skip therapeutic measures that would have cured the affliction, but drastically and perhaps irreparably slowed his progress toward the yokozuna rank. Despite all the surprises of these lean days, a close friend and confidant was able to say of him, "Jesse is a good-natured man, and that contributed tremendously to avoiding ridicule or criticism at the beginning."

Nevertheless, criticism may have affected him. Though he did outstandingly at the lowest Sumo levels, the climb up the Sumo pyramid was not easy. In 1966 alone he suffered three consecutive tournament losses, dropping Jesse precipitously down the Sumo ladder. Yet he persevered and rose steadily, if not speedily, since then. Around 1968, just four years after arriving in Japan, he reached the much-sought-after *Maku-uchi* rank, often regarded as the pivotal point for a professional wrestler.

There is a remarkable incident connected with these early phases in Jesse's career. Proof positive of the surprising Japanese openness toward foreigners is his surprising win of the *Kanto-sho*, or "Fighting Spirit" award, given in virtue of his 9-6 score in the 1968 New Year Grand Tournament. Though two other wrestlers surpassed him in actual wins, all concerned agreed that the young Takamiyama fully deserved the award, in view of his extreme problems in adjusting to Japanese life. To date, the high point of his career came when he captured the championship Emperor's Cup at the Nagoya Summer Tournament. That, plus his having competed in over 60 consecutive tourneys in the very top class, stamps him as one of the all-time greats, surpassed perhaps only by the immortal Taiho. Though he has been in the game for some 16 years, he remains a feared opponent whenever he competes.

His training is, by his own account, far more rigorous than that of other Sumatori. "So strict was the training, that in the first year I wanted to go back home. But . . . I have made good in the sport because my people in Maui, and the rest of Hawaii, have kept encouraging me. I stamp my feet 500 times a day, ten times more than other wrestlers, besides doing all other kinds of regular exercises. If others train 30 minutes, I practice twice as long." Like so many other strength athletes, Sumatori warriors have recently begun to practice systematic weight training and it is rumored that the great Takamiyama is capable of bench presses well in excess of 500 pounds, as his "pushing" power is by now legendary. With serious specialization, there is no question he could approach the world records in powerlifting. (The great Kaw-shiwado, also, was known to have been an ardent practitioner of the Iron Game as far back as the late '60s.)

By our standards, Jesse's diet would seem designed to drown a bull elephant. It is reputed that the Hawaiian superman has downed as many as 15 bottles of Japanese rice beer at once, though, like the Scottish strongman Bill Anderson, he shies away from the hard stuff. With a bodyweight in excess of 400 pounds at times, the great Takamiyama knows well the role of size and power in the Sumo arena. Megacarbohydrates are the order of the day. Perhaps surprisingly, Jesse has no fear of the much-touted "early-death" syndrome that the uninitiated make so much of when men of Takamiyama and Alexeev's girth are the subject of conversation. And Takamiyama's fearless attitude does have a solid basis in fact. All statistics show that despite recent popular magazine reports to the contrary, there is no difference whatsoever between the life spans of Sumatori and the general population. One possible explanation is that with the likes of Jesse and Alexeev, the percentage of bodyweight that is actually fat is far smaller than one might think. Though not abundantly in evidence, Alexeev and the Sumatori are mostly muscle, an active, healthy tissue that puts no added strain on the heart. It is known, anyway, that most of the old-time strongmen, wrestlers, Sumatori, etc., lived to ripe old ages. Russian wrestler George Hackenschmidt was having philosophical conversations with the editors of *The Humanist* when well into his 80's. Such facts ought to forever put to rest the myths about strongmen dying young.

Nor is that myth the only one Jesse Kuhaulua has squelched. The charismatic "blubber-tosser" has convincingly exploded the notion that all strongmen are musclebound. In recent months he has been seen on TV commercials, gaily decked out in 1930s-style pinstripe suits, flowers, and—wonder of wonders—tap-dance shoes! While twirling a yo-yo with considerable skill, he does a softshoe that would make Fred Astaire green (well, pale green, anyway). One is reminded of Franco Columbu's marvelous scene in *Pumping Iron*, where he does a bag-punching, rope-skipping routine with blurring speed, despite 187 pounds of muscle on a 5-foot, 5-inch frame.

Finally, the great Takamiyama is beloved all over Japan and Hawaii as a dedicated family man, benefactor of innumerable charitable causes, and universal lover of mankind. One of the most poignant, amusing, and delightful sights in all Japan—seen only by the most fortunate—is that of a multitude of small children struggling mightily to topple his enormous frame. One is reminded of Dicken's marvelous remark about the Crachit children in *A Christmas Carol*.

"The noise in the room was perfectly tumultuous, for there were more children there than Scrooge in his agitated state of mind could count; and, unlike the celebrated herd in the poem, they were not forty children conducting themselves like one, but every child was conducting itself like forty. The consequences were uproarious beyond belief; but no one seemed to care; on the contrary, the mother and daughter laughed heartily . . . and the latter, soon beginning to mingle in the sports, got pillaged by the young brigands most ruthlessly."

5

The Most Ancient
Art of Wrestling

Where is there a sport as interwoven with cultural history, intellectual concerns, or theological doctrine as wrestling? The Bible itself, in the Epistle to the Ephesians (6:10), mentions the heavenly combatant prepared to wrestle Satan: "Our wrestling is not against flesh and blood but . . . against the world-rulers of this darkness . . . Stand therefore, having girded your loins with truth."

Far from being a pastime of the untutored masses, the sport has mesmerized the noblest and most philosophical of men. It can safely be asserted that wrestling is the only sport that played a role in the development and growth of a major philosophical worldview. Aristotle, the greatest philosopher of ancient times, mentions the wrestler Milo in his *Nichomachaen Ethics* while explaining his doctrine of the Golden Mean. The philosopher's point is that in all things in life we must be careful to find the middle path—neither too much nor too little of a good thing. It is indicative of the high place the ancients reserved for wrestling that Aristotle would dignify so austere a theory by taking his example from the "Grappler's Game."

Many of the great intellects in history were wrestling fans. Abe Lincoln, a sinewy and hugely powerful man, was ranked among the best in the United States. In his eighteenth year he soundly whipped the powerful Louisiana grappler Jack Armstrong. During his tenure as president, wrestling became an integral facet of military training and is still emphasized at military academies today. Jackson, Grant, Taylor, Teddy Roosevelt, and Taft were all wrestling buffs. Taft is said to have thrown political opponents in more than one manner, by virtue of his great strength. In Japan, the Sumo titans are regarded as much philosophers as warriors, so densely overgrown is Sumo with the lore, mysticism, and eternal truths of the Orient. Considering that the East has always been a fruitful ground for philosophic inquiry, it is not surprising that

53

Wrestling may be the most basic form of combat. (Office of Public Information, Lehigh University)

Japan's national sport is so related to profound thought, especially when one considers that Japanese wrestling goes far back into antiquity, predating even the ancient Greek practice of the sport.

Yet wrestling, in other forms, can be traced back even farther. Indeed, it is likely that wrestling has been practiced ever since the first Neanderthals emerged from their apelike ancestors. It may be the most basic form of combat, related, as it once was, to the very survival of the species. For it is one of the few strength sports that needs no tools or implements whatsoever, save the strength and cunning of the combatant. Furthermore, it may well have been, at the dawn of *Homo Sapiens*, the only way to demonstrate male superiority. And despite the rapid evolution of weaponry in today's world, the combat soldier can find himself forced to rely upon this most primitive form of combat for his survival.

Perhaps only boxing compares in terms of its basic nature, yet we find no records indicating that boxing played any great role in the survival of primitive cultures. Nor is this unexpected, inasmuch as the fighting superiority of the wrestler over the boxer seems indisputable. Whenever a wrestler has met a boxer, the boxer has always lost. In 1887, for example, the fabulous John L. Sullivan opposed the great wrestling hero Bill Muldoon. Within seconds Muldoon had Sullivan in enormous difficulties, picking him up bodily and slamming him to the canvas with awesome force. Some years later a heavyweight boxing champion named Bob Fitzsimmons boasted proudly of what he planned for the Greco-Roman wrestler Ernest Roeber. Yet when the smoke cleared, Fitzsimmons was on the canvas, wrapped up most thoroughly in an arm-lock that Roeber held until he felt Fitzsimmons' humiliation to be complete. Indeed, the heavyweight boxing champion Jim Corbett himself declared that it was virtually impossible for a boxing champion to defeat a wrestler of comparable abilities.

It is around the time of the Egyptian dynasties, some 5000 years ago, that we find the earliest clear proofs of the practice of wrestling. It was also during the great dynasties that we find the most rapid advancements taking place in technique and skill. Many of today's modern forms of wrestling differ little from the sport as it was practiced in Egyptian and Assyrian times. And, like the discus, shotput, and other field sports, evidence of the existence of wrestling has been discovered on many ordinary implements of ancient days. There exists a statue, found in Mesopotamia, showing two wrestlers in combat. There is a wealth of paintings found in the temple of Beni Hasa on the Nile that clearly depicts wrestling holds, locks, and stances. There is one, for example, showing a wrestler locking the opponent around the waist in a way reminiscent of modern Cumberland-style and free-style wrestling.

Following the collapse of the great dynasties, interest in wrestling passed to ancient Greece. Both the *Iliad* and the *Odyssey* of Homer mention the sport, and back as far as 704 BC it had become an integral part of the Olympic Games. Greek legend and history are filled with stories of great wrestling matches, matches that not only validate the claimed importance of wrestling in the ancient city-states of Athens and Sparta, but clearly demonstrate the integral connection between the world of wrestling and the world of superhuman strength. Among the most widely known of these is the celebrated match between Hercules and Antaeus. Since his mother, the Earth, regenerated Antaeus' strength every time Hercules dashed him to the ground, Hercules finally defeated Antaeus by hoisting him over his head and breaking his back. (In the Joseph E. Levine film of the early '60s, Hercules was played by the immortal bodybuilding superstar Steve Reeves, with wrestling and boxing star Primo Carnera playing the role of Antaeus.)

The ancient hero Theseus wrestled the minotaur in the palace of King Minos of Crete, and also defeated the evil Cercyon of mythology. As the story goes, this ancient monarch challenged and defeated all who passed through Eleusis. When Theseus entered Eleusis, he promptly encountered Cercyon and just as quickly disposed of him with his overwhelming power and skill. Legend abounds with similar stories, including the apocalyptic battles between Dares and Engellus, Hercules and Achelous, and the epic struggles of Ajax and Ulysses.

But the most famous ancient wrestler was probably Milo, who had many unbelievable stories attributed to him. One holds that Milo developed his awesome power by carrying a calf on his shoulders every day until it was full-grown. Even then, so the story goes, he stopped only because the bull was growing too unruly and too irksome to bother with. Whether true or not, Milo is known to have won championships in the Olympic Games for 20 years, ending his reign in 516 BC. He captured the laurels in the games at Pythia, Nemea, and Isthia. There is one quirky story suggesting that Milo and Lysander battled to a stand-off for hours until Lysander finally dropped dead. Oddly, the victory branches went to Lysander, on the grounds that Death, rather than Milo, had won the day!

There is conflicting evidence as to how matches were conducted in ancient civilizations. Though it is commonly agreed that wrestlers oiled their bodies before a match, no one is quite sure why. (One explanation is that doing so made it more difficult to get a grip.) As practiced in the Pancratium of 704 BC, the first Olympiad of ancient times, the fights were conducted naked and the only rule was to gain victory in whatever manner possible. Brutal, savage, uncompromising, the bouts allowed virtually every trick in or out of the book and, not surprisingly, often ended with the death of a wrestler. Fortunately, however, the period of such unprincipled matches was relatively short. With the conquest of Greece by Rome, the sport became subtler and more sportsmanlike. The cruelest and most savage maneuvers were outlawed, and it became similar to what we now know as wrestling proper (as opposed to "pro" wrestling, which, despite its silliness, captures in some ways the essence of ancient wrestling with its savage facade).

All this is not to suggest that wrestling existed solely in Egypt, Greece, and Rome. A vast number of cultures, both ancient and modern, have told tales of heroic figures in the guise of wrestlers. The epic of *Gilgamesh*, for example, tells of Gilgamesh, the wrestler in the ancient land of Sumeria. According to the story, a powerful wrestler named Enkidu was created by the god Aruru to check the lustful excesses and ravages of an unrestrainable Gilgamesh. One night Gilgamesh, during a raucous party, entered a house of "ill repute," as he was wont to do. Standing before the door, however, was the newly-created Enkidu, hoping to turn Gilgamesh away from the path of evil. Furious, Gilgamesh hurled himself at Enkidu, delivering blow upon blow, thinking to best his foe as he had so many others. Yet Enkidu was quite the match for Gilgamesh and, try though he did, he could not overcome Enkidu. The struggle continued unabated, with an eager crowd awaiting the result. Abruptly, presumably realizing that Enkidu was not to be vanquished, Gilgamesh released his grip, threw back his head, and lauged uproariously. He and Enkidu became close friends thereafter, thus vindicating an earlier prediction of the prophet Ninmah that from the battle good would eventually come.

Centuries later, wrestling spread over the continent, eventually becoming a sport of royalty. (In the late 12th and 13th centuries, an overpowering warrior or monarch was thought to be living proof of the greatness of his reign and of his people.) The British Isles, for example, developed brands of wrestling uniquely their own, though their influence reached to other lands and other peoples. It is alleged that an epic battle took place in the 16th century between Henry VIII and Francis I of France, though the story is not universally accepted. One account asserts that Henry, for no apparent reason other than whimsy, decided to match his prowess against the considerably weaker Francis during the festival of the Field of the Cloth of Gold. Royalty and machismo being what they were, the French monarch could hardly refuse. Surprisingly, though Henry far exceeded him in raw physical power, Francis cunningly turned the tables by diving at Henry's bad leg, unbalancing him, and subsequently pinning him.

This famous story is occasionally cited as evidence of Francis' lack of sportsmanship. Yet the interpretation seems quite unfair to the French monarch, considering the circumstances and the cultural history of wrestling. Even ignoring the fact that Henry deliberately started the fight, intending to humiliate a weaker opponent, it must be noted that in all times wrestlers have regarded the sport as a free-for-all. Because the earliest cultures considered wrestling the most fundamental expression of good vs. evil in the human soul, it was thought limiting to place any restraints on the manner of conducting so basic a combat. Today, of course, the rules of wrestling—which holds are allowable and which are not—have been precisely defined. Even so, many cultures permit an astonishingly free approach. The *tsuppari* thrusts in Sumo, delivered bare-handed, strike the uninitiated Western observer as potentially more lethal than the best of Ali, though there are far fewer injuries in that sport than in many Western ones.

Nor do the great Cumberland wrestlers of the Scottish highlands "tiptoe round the tulips" either. According to the renowned bodybuilding and Scottish Games authority David Webster, many of the best wrestlers in the world have emerged from Scotland and England. As Webster explains in his *Scottish Highland Games*, "In the Scottish border country and the northern counties of England some of the world's best wrestlers have been bred, and the gatherings have long profited by the stern competition from stalwarts of these parts."

Webster tells an interesting story which may help to combat some of the nonsense one often hears from martial arts adherents who assume that a good karate or judo expert is bound to overcome "musclebound" wrestlers and strongmen. On the occasion of the Mintlaw Games, a group of Japanese wrestlers issued an open challenge to all comers in Great Britain. But, as Webster explains, "Unfortunately for the Oriental, in the crowd that day was John V. Gray, a farmer from Old Don. John took hold . . . like a dog with a ferret, to use the description of a reporter present at the match, and in half a minute—yes, thirty short seconds—he had the squirming Judo expert firmly pinned." Stern warning to martial-arts types. Yet it is not theoretically surprising. Matched against strength specialists—bodybuilders, weightlifters, or wrestlers—martial-arts men, though skillful in their sport, find the strength difference too much to overcome.

Be that as it may, wrestling began to be included in the modern Scottish Highland Games in the late 19th century. The old severe restrictions on the use of hands and arms were discarded and the so-called "Scottish," or "Donald Dinnie" style (named after the great Scottish champion who advocated that approach), emerged. In this form even greater than normal strength was required to win a bout, as victory demanded that the vanquisher hold his opponent down for a full half-minute. Eventually the Scottish style evolved into a kind of "catch-as-catch-can" style, similar to the old Abe Lincoln version of the Grappling Game. This demanded that a fall be scored only when both shoulders touched the ground simultaneously. Both it and the Cumberland style have remained essentially unchanged.

In still one more version of the British approach, Cornish and Devon wrestling demand that three, rather than two, points of the body touch the ground for victory (two shoulders and one hip, for example.)

Nor was the American Indian unacquainted with wrestling. Long before the invasion by the Europeans, legend and cave drawings tell of fierce matches between brave warriors. The epic battles one hears of between the Indians and the settlers with deadlier machinery often obscure the friendlier wrestling matches the pioneers engaged in with their Indian brothers. So influential were the Indians on the pioneers that a fairly large and organized network of matches and champions had developed in America by the end of the 19th century.

Worthy of mention is the legendary Martin "Farmer" Burns of Iowa. Though small, Burns was massively strong, and he brought the techniques of catch-as-catch-can wrestling to a new level of sophistication. For a while he held his own against the best, winning the pro wrestling championships in 1895 by soundly whipping "Strangler" Lewis. A few years later, however, both Tom Jenkins and Dan McLeod succeeded in defeating him. Only slightly larger than Burns, Tom Jenkins became American Champion by the close of the 19th century, holding that position until his defeat by the great Frank Gotch in the opening days of the 20th century.

One of the all-time greats—perhaps *the* greatest wrestler of all time—was the Russian superman George Hackenschmidt. Aptly nicknamed the "Russian Lion," Hackenschmidt was not only enormously powerful—one of the very mightiest—but quite a philosopher as well. According to a story told to me by Lloyd Morain, editor of *The Humanist*, Morain was called to a social gathering in the '50s by his friend Ray Van Cleef of *Strength & Health* magazine. Unbeknownst to the genial editor of *The Humanist*, there was Hackenschmidt in his carefree 80's, anxious to discuss philosophical matters. How competent a dialectitian Hackenschmidt was, Morain did not reveal. Yet in the world of strength Hackenschmidt was without peer. Though twice defeated by a recognized inferior, Frank Gotch, most experts concede the superiority of the great Russian lion. Certainly he performed feats of strength that Gotch never matched, and had no difficulty whatsoever against Tom Jenkins, a distinguished wrestler and one who gave Gotch considerable difficulty. According to a newspaper report, "Jenkins was handled like a pigmy in the hands of a giant . . ." Also, Hackenschmidt easily handled the terrible Turk, Ahmed Madrali, in a match that lasted a scant 90 seconds.

Wrestling in the United States took a rather kaleidoscopic course, changing hands so rapidly even the papers of the day found it difficult to keep track. From 1914 to 1917, Henry Ordeman, Charles Cutler, Benjamin Roller, Joe Stecher, and Earl Caddock all garnered the laurel wreaths of the game. Still, many observers felt that Stecher was the preeminent postwar combatant, though in my judgment "Strangler" Lewis (no connection with the aforementioned wrestler of the same name) was his equal. Indeed, Lewis defeated Stecher in 1920 and held on until 1925, with only one brief loss of the crown, when Wayne "Big" Munn soundly thrashed him in Kansas City.

After Strangler Lewis won back his title for the last time from Stecher in 1928, the era of real professional wrestling came to an end in the United States. Interestingly, the first of the show-biz tactics that led to the hideous distortions of legitimate wrestling we see today was invented by Gus Sonnenberg of Dartmouth. A capable and powerful athlete in his own right (he beat Lewis in 1929) he was also a skilled businessman and showman. His so-called "flying tackle" spawned the tactics of the likes of Freddie Blassie, the "Executioners," and other throwbacks to medieval morality plays. Historically, one of the most famous of these clowns was "Gorgeous George"—a most entertaining monster-pixie with bleached-blond hair, best known perhaps for his famous bouts with the awesomely powerful Bob Hope. Planned and wonderfully choreographed down to the last detail, this sort of thing has nothing whatsoever to do with real wrestling.

Genuine wrestling in the country began around the end of the 19th century, when the AAU took authority. The style of the time in the United States was freestyle, though other countries, such as Japan and Scotland, maintained their time-honored styles. Freestyle and Greco-Roman jockeyed for preeminence for several years on the Olympic schedule until, in 1920, both approaches became permanent features of the Olympiad. The organizational structure of the sport remained relatively unchanged until 1973, when the *United States Wrestling Federation* superceded the AAU as far as international competition was concerned.

As practiced today, freestyle wrestling involves a rather liberal use of the lower body. Matches take place on a mat a little over 39 feet square, the actual contest circle being a bit over 29 feet in diameter. A fall is scored when both shoulders are held to the mat for a full three seconds. Unlike their more savage predecessors, today's amateur wrestlers cannot force a knee or elbow into the abdomen, employ a double-armed headlock, or bend the opponent's arm more then 90 degrees, among other things. This makes for triumphs of strength and strategy without excessive pain or injury.

In the Greco-Roman style (which, as indicated earlier, may bear little resemblance to the style actually employed in Plato's time) the rules and conditions are more or less the same, though use of the legs is severely restricted. Gripping the opponent with the legs or grabbing his legs are not allowed. The name of the game is aggression; too passive a stance by any participant results in penalties, as do arguing and fouling.

Predictably, training for top-flight competitive wrestling is intense. High-protein diets, weight training, and, especially, a bewildering variety of cardio-vascular exercises supplement most training programs. So basic is wrestling, however, that eventually the most dependable rung on the ladder of improvement is simply more wrestling, followed— guess what?—by more wrestling! In this sense, improvement in wrestling parallels improvement in the squat or gaining fluency in Chinese. The only way to get better is by doing it. And there's nobody in the world today that does it better than Lehigh's famed Mark Lieberman.

Mark Lieberman

At 25, Mark Lieberman is virtually an infant in big-time wrestling. His Bible-thumping nemesis, John Peterson, is several years older, as are a number of other top-ranked rivals. And he was late in starting as well.

The road hasn't always been the smoothest, as when big John Peterson attracted the Almighty's attention just long enough for Him to help Big John squeeze out a 7-3 victory over Lieberman in 1979. Still, Lieberman is stoical. "If it were easy, I wouldn't consider it worth doing."

The setbacks have been few, however. In his career, Lieberman has piled up an impressive *curriculum vitae*. Twice the NCAA champion, twice the *United States Wrestling Federation* (USWF) champion, and once the National AAU champion, he has rolled up an awesome 85-6 record, with 43 falls, during his tenure at Lehigh alone. He has won his last 47 collegiate bouts in a row, was the gold medalist in the Pan-Am games in Mexico City, the World Cup winner, and a four-time *Eastern Intercollegiate Wrestling Association* (EIWA) champion—only the third in history to accomplish the latter.

On top of all that muscling out of his opponents, he's completed a double major in accounting and international relations, compiled a powerful academic record, and munches stale chocolate at Helen's Lunch while other competitors juice up on anabolic steroids. Still, they say Lieberman has no talent. Lieberman himself remarks, "My power isn't great, my speed isn't great, my balance isn't particularly good." When asked where the heck his victories come from, he answers, "I beat people in the mind." Tom Hutchinson, of Blair Academy in New Jersey, who groomed Lieberman for Lehigh, also says, "In seven years here, I've had at least fifteen wrestlers with better athletic ability."

Is all this credible? How in the hell do you beat somebody *only* "in the mind?" Lots of guys like to say this—I heard some disgruntled weightlifters say the same thing when the colossal Alexeev blasted his latest world-record weight overhead. Inferior bodybuilders complained that the great Schwarzenegger used to "psych them out" in competition. Even the "Oak" himself said that. Excuses galore.

Lots of people have the right mental attitude. Bruce Wilhelm, the American weightlifter, had the right mental attitude. George Burns has the right mental attitude. *I* have the right mental attitude. But Alexeev and Rakhmanov probably don't juice up on vodka worrying about the three of us. Winners win because they're better than anyone else, plain and simple. These sorts of attempts to analyze losers, to collapse athletic ability into "psyche," "attitude" and the like are, beyond a certain point, journalistic fodder. As any good Skinnerian behaviorist knows, the only meaningful measure of talent is performance.

All this is not to deny that things like attitude and hard work are components of a champion's make-up. Lieberman himself used to say, "I have left the wrestling room with blood running down my face, tears streaming down my cheecks, soaking wet, and I can't help it. I just sit down and have a good cry. Then I go back into the room and get pummeled again. It's that kind of sport." This uncompromising approach doesn't fail to rub off, either. A promising 134-pound wrestler, Darryl Burley, also of Lehigh, says of Lieberman, "He tells me I can beat people who are better than I am because I love it more. Like him." Yes, Virginia, there is room for psychology. But only a little.

Even losses figure into Lieberman's total program. "If you let losses beat you," he argues, "you're beating youself. I think it's important that people know you fail. It shows you're human. When you hide your losses, it makes you afraid to lose. Frankly, my goal is to lose 1000 times,

Mark Lieberman, champion wrestler. (Office of Public Information, Lehigh University)

because I know that in order to lose that much, I'm going to have to wrestle anytime, anywhere, against anybody. . ."

Sociologists and philosphers from Durkheim to Glazer have told us how a cultural environment shapes the ultimate being. And they'll not find contradiction in Lieberman. Born to a family destined for success, his brother Mike was himself an NCAA champion. Papa Lieberman is a highly-ranked executive for Lieberman-Harrison, Inc., an advertising agency in Allentown, Pa. and New York. A driving, relentless force in the molding of his children, he nevertheless mixes apparent ruthlessness with stoic calm: "I always insisted that the kids give anything a real hard look before they started, because once they did, they knew it had to be completed in some reasonable fashion."

Nor did having big brother Mike in the limelight make things any easier for Mark. Three years ahead of Mark, Mike had every reason to believe he was the darling of the wrestling gods—heir apparent to the crown of Hercules and Theseus. The stellar accomplishments of an older sibling have driven more than one potentially star-bound individual into a web of hopelessness. Not Mark. His brother's virtuosity on the mat served only to drive him to loftier heights. That kind of relentlessness and determination paid off. In the 1976 Olympic regional trials Mark triumphed over big brother Mike by a score of 5-3. Later, in the national trials, he squeaked by big Mike by one pin. Mark's attitude? "It wasn't much of an up. He's an idol, and you don't like to see an idol tumble. Especially, you don't like to tumble your own idol."

The final chapter in Mark's encyclopedia of accomplishments remains to be written. Numerous hurdles intrude. For one, there is an ever-increasing number of world-class athletes in feverish pursuit. Peterson is good and always dangerous. The Soviets obviously can't be overlooked. Also, Lieberman's career has been littered with injuries. In his freshman year at Lehigh he smashed tissues in both knees, topping that off with a serious injury to the right arm. The wounds healed quickly, but Mother Nature will tolerate insult only so long, and many great careers have been laid to eternal rest by seemingly routine mishaps. The great John Davis permanently ended his career with a thigh injury; the same fate may await the great Alexeev even as pundits speculate about the unthinkable 600-pound clean and jerk. The great Sumo warrior Tamanoumi yielded to the Fates after illnesses wracked his once indestructable body. The list is endless. Still, Mark Lieberman's stoicism would have made Epicurus proud: "Your body pays for living. It's not built for abuse, and we abuse it every day. We live a risky life. If you're going to play, you're going to play injured. The main thing is that you can't let it dim your competitive instinct."

The future? Who knows? At 25 (as of this printing) Mark is young enough to show he could be the best there ever was. Maybe then they'll think he's got talent.

6

Shotput: Explosions of Power

Shotputting—the darling of all muscle track-and-field events, sport of lords and warriors—may be the most basic form of weight-throwing. Like so many other strength competitions, its roots can be traced so far back into antiquity that it is difficult to assign a precise date of origin. Stone "putting" is mentioned in the Book of Leinster as an event in the Irish Tailtin Games in 1829 BC. In his famous mystical poem, "The Lady of the Lake," Sir Walter Scott tells how the awesome Douglas defeated all comers in the sport quite handily.

It is likely that the Greeks indulged in some form of stone throwing in accord with Socratic admonitions to develop the body equally with the mind (though there is no evidence that it was as widely practiced as throwing the discus). By medieval times, weight-throwing had gained enormous popularity in England—so much so that Edward III forbade "such frivolous events for fear that the sport could come to overshadow archery." Naturally, the good king could not allow this, as archery was far more relevant to matters of war than stone putting. Later, however, in the reign of Henry VII, himself a great fan of field events, the tradition of "casting the barre" became a regular daily exercise of his and was encouraged among the gentry. As modern methods of warfare developed, enthusiasts began using spare cannonballs instead of rocks for throwing.

Regulation size 16-pound shots were adopted much later on, in the late 19th century, at Dublin University as well as at Oxford and Cambridge. Sheer muscle power was the sole ingredient in the early days of modern shotputting. Any advantages in technique one competitor had over another were inevitably a matter of chance, scientific throwing methods not yet having been conceived. Still, a marvelous example of one who did exhibit both power and form was the great Ralph Rose of California. In the opening days of the 20th century, Rose combined overpowering force of limb with a style that may have rivalled that of contemporary athletes. He heaved the iron ball 51 feet—a world mark that is credible today. (Dave Pellegrini of Princeton, for example, boasts a personal best on only 49-plus in the event, though he is one of today's better hammer throwers.)

In fairness, it must be noted that occasionally a smaller man with virtuoso technique did best the giants of the day. Clarence "Bud" Houser won both the shotput and the discus throw in the Antwerp

Putting the shot as Al Feuerbach does calls for an enormous surge of power. (Emporia State Photo Services)

Olympiad, though many of the other competitors outweighed him by 100 pounds. By and large, however, the shot is an event for the big man. Technique and training methods have become standardized to the point that reliance on technique alone will not bring home the medals. The mighty Jack Torrance set the world mark in 1934 with a toss of 57-plus feet, while weighing a mind-boggling 275 pounds and standing nearly 6½-feet high. So overwhelming was his size and brawn that his nearest rival, John Lyman, was unable to ever really come close, weighing a "mere" 200 pounds.

As with so many of the strength events in this book, there would be a colossal gap without some mention of that "grandaddy" of most modern strength events, the Scottish Highland Games. Throwing the stone was a widely practiced activity in farms and villages in agrarian Scotland, primarily because finding the implements was trivial. According to the world authority on the Scottish Games, David Webster, "For a long time, stone putting was the most popular informal activity of an athletic nature practiced in the farmyards and villages, and the reason for this is not hard to understand. The only equipment necessary was easily obtained without cost, and even the roughest ground is quite suitable for practice."

Indeed, the lifting as well as throwing of heavy stones figures prominently in the Scottish traditions of strength. The *clach neairt*, or "stones of strength," were often seen at the doors of Highland chieftains. (Even today similar stones can be seen in front of Scottish mansions.) In essence, such boulders were a sort of welcoming ritual: Any visitor able to hoist the gargantuan rocks proved he was of genuine Scottish blood. Webster indicates that such stones were also placed outside the churches of old Scotland, so that a young stalwart might demonstrate his prowess after blessings from the Almighty.

Apparently, the rules for throwing the old Scottish shot varied quite a bit from the standard, well-defined rules of today. In some ancient drawings, athletes employ a sort of "overarm" style that obviously would produce much larger distances than today's methods. Also, in the earliest practice the throwers began from a certain mark, but were disqualified if they left the defined boundaries before the stone was released. Once the shot was thrown, though, going several feet beyond the boundary was quite legal. No such crossing is allowed today.

As far as technique is concerned, Webster finds it erratic and haphazard. He argues, "Raw novices and many of the biggest men, too, propel the shot into the air with the power of arm and shoulders alone, but experts like Arthur Rowe have the ability to utilize most of the muscles of the body." From what is known about the science of weight-throwing today, this is correct. The best shotputters, discus, hammer, and javelin throwers use the large muscles of the legs as a major part of the explosive power needed to reach world-class distances.

How have the Americans done in the great games of Scotland? Reasonably well, though they have hardly overwhelmed the Scotsmen despite the technique and weight-trained muscle of the USA's best. Brian Oldfield did create two records in 1973, but the burly iron-tosser was overwhelmed by the Scottish superman Bill Anderson in the overall tally. Indeed, even the records of Scottish greats of many years past compare favorably with the marks of today. Donald Dinnie, that archetypal Scottish gamesman, threw the stone over 45 feet in 1867 though even then he was starting to decline. Some experts claim that Dinnie's performance on that occasion (with a stone weighing 20 pounds) would be the equivalent of a toss of about 65 feet today. Such a heave would be within reach of world-record distances and far in excess of all but the best collegiate throws of today. (Is there magic in the Scottish diet of porridge and kippered herring?)

Still, Americans have shown improvement in shotputting through the century. In 1948, Charles Fonville set a world record with a toss of 58 feet. Though somewhat diminutive by today's standards (194 pounds), he more than compensated for this with his overwhelming technical prowess and blistering speed. From that point on, the records fell like so many dominoes. The great Parry O'Brien dominated the scene in the '50s, until displaced by the marvelously muscled Bill Nieder. In 1960 Nieder, at 242 pounds, reaffirmed the advantages of bulk by hurling the iron sphere a smashing 64 feet, 6¾ inches. Yet the sleeping giants in Great Britain could wait no longer to remind the world of their historic domination of the sport. In the summer of 1961, Arthur Rowe came within biting distance of the world mark by heaving the shot a full 64-plus feet.

That signaled an historic development. I believe it safe to say that the advantages of modern systematic weight training were beginning to make themselves felt. Though weightlifting was put down by most "authorities" as late as the early '50s, the great Arthur Rowe was including heavy lifting in his routines by the '60s. In 1961, Rowe was capable of bench pressing over 400 pounds and squatting with over 600 pounds. Still another weight-trained athlete, Randy Matson, continued the assault on the best distances with an unbelievable toss of nearly 72 feet in 1967. Even today this is startlingly close to the world mark of 72 feet, 8 inches, set by the great German Udo Beyer in 1978.

Is there a limit to the ultimate distances some future superman might attain in this event? Historically, attempts at forecasting limits in the strength events have rebounded ludicrously. In fact there are a number of athletes who may very well raise the mark to a staggering 75 feet by the end of the decade. And when one speaks of the gods of weight-throwing, one must revere the name of Udo Beyer.

Udo Beyer

"I was so surprised at winning, that a cold chill ran down my spine."

Vintage Beyer—a friendly, unpretentious giant who has almost competely dominated the sport of shotputting since his resounding victory in Montreal. However, the great East German is not ready to rest on his laurels. As he explains, "It is clear that I then wanted to show that my Olympic victory was no flash in the pan and to take away the tone of fortuitousness from it by putting up good performances later. I'm still in that position today and shall not be out of it until I have been able to gain a place well up front at the 1980 Olympics in Moscow."

Udo Beyer, Olympic winner in the shotput. (ADN-Zentralbild)

Admittedly, the triumph at Montreal was something of an upset. He was hardly a household word even in the GDR (German Democratic Republic) prior to the Olympic victory. He was not the national champion and the decision to include him on the team was made at the last possible minute. Perhaps to prove that the world-record performance was hardly a fluke, the powerful East German has placed even greater burdens on himself since the Montreal Olympiad. Only the '80 Olympics is not in his pocket, when Kiselyou of Russia took the gold.

Beyer took the laurels at the European and World Cups in 1977 and 1979 and, at the International Track and Field Championships in Göteborg, Sweden, he upped the world record with a staggering 22.15-meter (72.67-foot) shotput—the first time in history anyone has broken through the 22-meter barrier. In the European Track and Field Championships in Prague, he rolled over the finest European athletes, thereby allowing the GDR to gain its fiftieth victory in that event. Even in triumph, however, the genial champion waxes critical. As he noted of his winning throw of 22-plus meters in Prague, "Half a meter more would have fitted both the title and my face better." Nor does he bother to mention that the damp, chilly weather and hostile conditions sliced into the performances of all the competitors.

His training borders on that of a religious zealot—hours and hours of weightlifting, technique, elasticity exercises, throwing, and running, regardless of weather. His diet is typically infused with massive amounts of protein and, most likely, anabolic steroids, all complemented with enough carbohydrates to rev up his energy system. Beyer absorbs the regimen philosophically, explaining, "I'm just mad about putting the shot and do everything for my sport. I do not want to disappoint anybody."

Desire is the distilled essence of Beyer's success. A devoted warrior of the shot, he is not yet oblivious to the larger universe around him. A devoted family man, he has a lovely wife, Rosemarie, and parents whom he counts among his biggest fans. Udo's three sisters and two brothers are all active athletes as well. Other cherished people in his life include Fritz Kühl, his trainer, and the faculty and students of the Leipzig College of Building, where he is currently studying Constructional Engineering. Additionally, his homeland recently compensated for earlier neglect (before the Montreal Olympiad) by voting him "Athlete of the Year."

Where does Beyer go from here? Despite 26 scant years and global preeminence, aspirants to the throne of shotputting hound him from every angle. But if his famed determination is any indication, he'll be around for decades to come.

Maren Seidler

Will massive weightlifter Gerd Bonk or shotputter Udo Beyer one day be on birth-control pills? For according to Maren Seidler, America's ace shotput star, that's exactly what the brawny *frauleins* from East Germany are using to gain muscle and surge far ahead of their American counter-parts. "I thought it sounded crazy," says Seidler, recounting her first conversations with the eminent West German coach Christian Gehrmann. The theory, according to Gehrmann, is that birth-control pills act similarly to the notorious anabolic steroids, increasing muscle mass and protein metabolism. Yet they remain legal—so far.

To date, Seidler has not touched either birth-control pills or anabolics. For this reason she has, despite her enormous natural strength and great talent, remained a step behind the Europeans. But she has every intention of closing the gap. Today she boasts of being a seven-time outdoor national champion and a three-time Olympic participant. She recently bettered her own women's indoor record (American) with a toss of 57 feet, ¼ inch in the '80 AAU indoor championships and she holds the outdoor record of 56 feet, 7 inches. At age 29 she is the "grand old lady" of American track and field.

Seidler's prowess in Olympic competition has been erratic: 9th in her first Olympiad, 14th in Munich, and 12th in Montreal. The obstacle to her ambitions has, for some time now, been the steamrolling Czechoslovakian Helena Fibingerova, who has set one world's record after another, and as early as 1977 had thrown nearly 15 feet farther than Seidler ever did. Still Seidler is philosophical: "I could be in Kansas City one day beating everyone by far, then go to Moscow and not even be in the running. Both ways there's no competition." In international meets with the Soviets, Seidler is the perennial number three. "I probably have a zillion third places," she will say wistfully.

Soviets and Czechs notwithstanding, the great Seidler pursues her goals doggedly and, while she did not get her chance in Moscow, will have ample opportunity to show her stuff in future meets. Perhaps the single greatest reason Seidler inches ever closer to world domination is that she, unlike so many competitors, has made a complete analysis of herself as a human being, an athlete, and a woman. She has the rare gift of being able to turn an apparent disadvantage into an asset. Her father, Walter Seidler, is 6 feet, 9 inches, and a former Long Island University basketballer. Maren, at over 6 feet, inherited his attributes. Knowing that for girls, especially, "being tall can be an awful thing," Papa Seidler told his daughters to "take pride in their height." Maren has.

Yet her original problem continues to dog her—she is too good for the Americans, yet unable to keep pace with the best Europeans. For this reason her motivation has suffered at times, a fact she is keenly aware of. According to Dr. C. Harmon Brown, coach at the Lions Track Club and her mentor for the first Olympic Games, "After 1968 she came home and discovered there was more to life than track and field. She is poised, bright, full of curiousity, a thinking person." Still, the lack of motivation is not her fault. According to ace javelin thrower Kathy Schmidt, "When it comes to women's field event training and competition, there is very little happening in the Unites States. I compete almost exclusively overseas."

The explanation is obvious enough. American society has traditionally put down the concept of women demonstrating athletic ability of any kind. As her friend and sometime coach Al Feuerbach, a former world-record holder in the shotput himself, explained, "I think that Maren has the talent to be an international class star . . . She does have the ability to be among the best." But he is quick to assert that American women shotputters have considerably more psychological hurdles to overcome than their East European colleagues.

Nonetheless, the peculiarities of the American psyche regarding women muscle-stars is changing rapidly. In recent years, both the sports of powerlifting and bodybuilding have been invaded by women. In powerlifting, both Ann Turbyne (herself a world-class shotputter) and Jan Todd have attracted global attention. Amazingly, these and many other stars have exploded the myth about "weaker" damsels waiting to be rescued without sacrificing a jot of femininity or natural intelligence and poise. Woman's bodybuilder Doris Barrilleaux has, at age 49, a physique that causes envy in women half her age while demonstrating that showy muscles need not detract from intelligence and charm. The change cannot but be beneficial for Seidler and her peers.

Even as the old prejudices vanish in the dust, there remains the intractable problem of the Communist system. Eastern Europe, which breeds top-notch athletes like flies, provides all promising younger athletes with total government support, a personal coach, and a program scientifically designed for each athlete. Seidler and other Americans are fighting, in short, an entire army of drug chemists, scientists, coaches, and a choking bureaucracy that makes it all but impossible for a lone American to compete on equal footing. By contrast, Seidler has had to fend pretty much for herself. She devises her training helter-skelter, depending on bits and pieces of advice from her friends, Mac Wilkins, Feuerbach, and others.

Still she persists. Since graduating from college, she has done nearly everything possible to negate the lack of an Eastern European type of environment. For starters she moved to the San Jose area, a mecca for track and field people in the United States. Happily, a copy clerk's job at the *San Jose Mercury News* gave her ample time to pursue the defiant metal sphere. Add to that an intensive weightlifting program that would crush most male enthusiasts. Three times a week she works with free weights to increase overall power and size, as well as explosive energy. As with all weight-throwers, she strives to maximize the force of the *latissimus* muscles of the back, as well as her triceps and shoulders. For overall leg power she employs leg-biceps exercises and heavy squats. The rest of the week she uses the controversial Nautilus machinery, though more for suppleness and general conditioning, as the machines are not known for building really enormous strength.

Seidler's world view is not as limited as many strength athletes. An anthropology major at Tufts, she has developed the intellectual vision to realize that the very best do not live in a galaxy of iron and medieval torture machines. She loves fine conversation, music, and has recently taken up meditation as well. Flamboyant, colorful, and engaging, the renowned amazon is not afraid to wear her identity in public. Often clad in green T-shirts, enormous triceps and biceps rippling, she sports gold earrings and silver bracelets in the manner of an Olympian warrior. Always philosophic and stoic in her eternal fight against female stereotypes, she sums up her attitude toward competition and feminity succinctly: "The dream of the typical American girl-next-door still doesn't include a good pair of quadriceps. To be a good shotputter, you can't be a tiny, tiny thing and throw far. But there's a big change in attitudes. Muscles are becoming OK for women; it's even OK to sweat."

Kathy Devine, an up-and-coming American shotputter. (Emporia State Photo Services)

How far can the artist of the shot go? At a recent NAAU outdoor championships, she upped her best distance to 62 feet, 7¾ inches. Thus, Maren is cautiously optimistic. Though reaching her ultimate will demand more competition from the world's best, strong challenges are erupting from many American athletes—Kathy Devine, Marcia Mecklenburg, and Ann Turbyne among them. In the interim, Seidler enjoys life, training and buddying with Feuerbach, Mac Wilkins, and Schmidt, and recalling her father's prophetic proclamation: "You see this little round ball? If you become good at this you'll be able to travel all over the world."

7

The Mystical Lore of the Discus

"Discobolus": possibly the single most revered, inspirational, and influential word in the history of weight-throwing. It is said that Myron's renowned statue, conjuring images of consummate grace, beauty and superhuman power, so struck the ancient Greeks that it was singlehandedly responsible for the birth and growth of the Olympiad.

Throwing the discus was a standard event at the very earliest games. Not only was it valued as a competitive sport, it was prized for its athletic and health benefits as well. Homer refers to the event in the *Odyssey*, ascribing the "world's record" to Ulysses. The philosopher Statius describes the practice also, though it is not known today to what extent the discus and the throwing of it resemble the contemporary phenomenon. Attempts at reconstructing Myron's "Discobolus" have been made, partly for historical and aesthetic interest and partly to obtain precisely the knowledge described above. There is such a statue in the British Museum, though wide disagreement exists as to the accuracy of size, shape, and the technique of tossing in that piece. According to many academic studies, the discus, or "quoit," may have weighed anywhere from 3 to 12 pounds, and the throwing style may have varied from running to stationary positions. In either case, however, scholars agree that the style differed markedly from today's widely practiced turning, or "corkscrew," style of throwing.

Interestingly, the discus was hardly the most popular athletic entry with many Hellenistic Greeks, despite its inclusion in the earliest Olympiads. The javelin was supreme, doubtless because the javelins or spears, like the log tosses of agrarian Scotland, played a direct and practical role in the lives of the early Greeks. Given the importance of the spear both in hunting and in actual warfare common among the rival city-states of the 5th century BC, it is natural that throwing the javelin should evolve into a ritual of the first importance. The only reason

Throwing the discus has a tradition going back to the first Greek Olympics. (Department of Intercollegiate Athletics, UCLA)

hurling the impractical discus evolved at all was because of the traditional Greek interest in aesthetic matters. Greek metaphysical systems, most importantly the powerful tomes of Plato, Aristotle, and Plotinus, are among the most influential writings in the West, and many of these works discuss artistic matters.

One measure of the importance of discus-throwing is evident in the inclusion of the *diskos* in the first Olympiad games of modern times, held in 1896. Naturally, the Greeks of the late 19th century regarded the occasion as an easy mop-up, given their historical acquaintance with classic throwing techniques. But the Americans jammed the machinery nicely. A little-known shotputter from Princeton, Robert Garrett, had begun training for the games with essentially no knowledge of either the construction of the discus or the proper techniques for tossing it. Oddly, this worked to his advantage. When he first tried a Greek version of the bulky platter, he realized it was considerably lighter than the homemade version he'd been practicing on. As a consequence, he sailed to victory with an unbeatable heave of over 95 feet.

Just one year later in the United States, Hennemann of Chicago won the first national AAU title with a fling of nearly 119 feet, inspired, no doubt, by the American victory at the first Olympiad. Hennemann's record stood for nearly four years. Centuries of Greek domination were over. The mantle passed to the Americans and was soon to land in the brawny grasp of an Irish-American policeman from New York City, Martin J. Sheridan, Olympic champion in 1904, 1906, 1908, national champion from 1904 to 1907, and again in 1911. He raised the earlier records to over 120 feet, finally heaving the discus 141 feet 4½ inches—a smashing throw, especially considering that Sheridan threw from the cramped confines of the then standard seven-foot circle. (It should also be noted that Sheridan set a record that still stands today—a throw of 124 feet, 8 inches in the freestyle, though the freestyle event hasn't been included in the Olympic games since 1908.)

At this point in the history of the discus, a revolutionary development occurred. Around the time that Sheridan set the last of his world marks, forces for change were beginning to assert themselves. Many began to realize that the seven-foot circle was far too confining, and that with the added leverage and momentum a larger circle would provide, the ultimate limits of man's discus-throwing ability could be fully explored. Finally, after considerable controversy, the new circle of 8 feet, 2½ inches was adopted for the national AAU championships in 1915. It is still in use today. It was also at that time that another of the greats of the game, A.W. Mucks of the University of Wisconsin, rose to establish a new niche in the halls of the immortals. Also, J. Duncan appeared to record a toss over 156 feet for still another world mark.

The new circle was proving itself. The number of participants in the discus swelled to even larger numbers. Nonetheless, many people continued to regard Duncan's toss as something of an oddity, never to be repeated. It did indeed stand for 13 years, until the intense competitive climate of California eventually pushed the mark to over 157 feet, set by Houser of USC. The latter, reminiscent of the great "Multiple" Mac Wilkins of today, also won the shotput in the 1924 Olympic games, coming back again to win the discus four years later. Houser's records were not destined to stand for long, however. Krenz of Stanford, Rasmus, and Jessup all conspired to force the record to a staggering 169-plus feet. By then the Americans were in full control of the discus. Many writers, however, attribute this to differences of technique and the inferiority of the European throwing style, rather than to talent, work, or anything of the sort.

For that reason a few remarks are in order: In the 9th Olympiad at Amsterdam, greats such as Hoffmeister, Handschen, and Paulus seemed overwhelmed, appearing as mere shadows of their previous greatness. Notables such as Marvlits, Egri, and Askildt seemed to have utterly lost their form and the Yankees rolled effortlessly and mercilessly over all the great competitors of Europe. Why? In the European style the arms described a wavelike rising and falling action. This produced an extremely uneven toss which sometimes yielded good throws but seemed not to be generally reliable.

The American style, by contrast, made no use of the wavy arm motion. W.E.B. Henderson spent considerable time developing a style similar to today's, where the throw is executed on a rising, nearly horizontal plane. The preliminary swings, according to Henderson, should be made from right hip to left shoulder across the body. This was to be followed by a kind of rising, corkscrew twisting of the body. Another difference from the early European style had to do with the method of supporting the discus. Like some renegade French throwers, the Americans often supported the discus on the palm of the hand (rather than in the European underhand style) while holding it in front of the body, incidentally helping to eliminate any trace of the notoriously inefficient wave motion of the European style.

Eventually, of course, the Europeans adopted the American style, so that today there is no difference whatsoever between accomplishments on either side of the Atlantic. If anything, the Europeans, with the appearance of female juggernauts such as Evelyn Schlaak of the German Democratic Republic, may soon lead the world in discus-throwing. Of course, any discussion of modern developments in the discus must mention such stirring performances as Fortune Gordien's mighty heave of nearly 200 feet in 1954. Unfortunately, this was disallowed as a world mark because the discus used was, inexplicably, one ounce lighter than the regulation four pounds, six-plus ounces. Had he used the correct discus, most experts feel he still would have set a world record.

Another bizarre incident occurred in the chaotic history of the quoit when, in a beautifully executed toss, Jay Silvester flung the platter 210 feet *downhill*. The throwing field actually fell off some 27 inches from start to finish. Why this site was chosen to execute one of the world's greatest fiascos has never been satisfactorily explained. Had the field been level, it is likely Silvester may have thrown a new record anyway, so great was his power and speed. Fortunately, he redeemed himself in 1968 with a world record of nearly 225 feet—stunning for the time, to be surpassed later only by the very greatest. And when we speak of the very greatest in the venerable history of the discus, we speak of only a very, very few—Oerter, Danek of Czechoslovakia, and one or two others. Yet at the summit of the pyramid stands the greatest of all: mighty "Multiple" Mac Wilkins of the United States.

The times being what they are, however, something needs to be said about the great throwers of the fairer sex. With the appearance of the dazzling Evelin Schlaak of East Germany and her record-smashing throw of over 226 feet in the 1976 Montreal Games, women are beginning to demonstrate their amazing capabilities. And with the relatively recent pursuit of systematic weight training by most female strength athletes, it is possible that they may soon equal or surpass the best men's performances. From one aspect, pure physical strength, athletes such as Jan Todd and Ann Turbyne are already stronger than 99 percent of the men in the world.

A proviso, however: For one, the discus used by women in official competition is only half the weight of the men's discus, making comparisons between the sexes tricky. Yet the size of the discus is not the only factor. It is conceivable that what women lack in relative size and strength, compared to their best male counterparts, could to a large degree be made up for by greater flexibility, leverage, and speed. And, considering the fact that since the introduction of the women's discus in 1928 at the Amsterdam games their records have increased at a considerably faster rate than the men's, we can expect the relative difference between men and women's performances to shrink ever more in the near future. Speaking of relative performances is one thing. The ultimate question, obviously, is whether women could equal men in *absolute* performances. Could a futuristic clone of Evelin Schlaak hurl the men's discus further than Mac Wilkins? The question admittedly borders on science fiction. However, anyone who has seen the great Ann Turbyne bench-pressing nearly 250 pounds knows she is within reach of 300 pounds. Now that's getting into respectable poundages, even for men! Thus, if a woman can bench-press almost as much as a man, why couldn't a woman shotputter or discus-thrower heave the men's implement within striking distance of good male performances?

For the present, we can only record historical accomplishments under conventional conditions and circumstances. With the women's discus, Helen Conopacka of Poland grabbed the Olympic crown in the Amsterdam games with a throw of nearly 130 feet. Her throw, though not spectacular by today's standards, was remarkable for the times. The next major advance in female prowess came in 1936 when the powerful Gisela Mauermayer of Germany catapulted the saucer a blistering 156 feet, 3 3/16 inches. Yet, following this, there was something of a decline; interest among women in the discus sagged, contributing to a relatively uninspiring throw of just over 137 feet by Ostermeyer of France in the 1948 Olympiad. The lull was short-lived, however, and in 1952 the Soviets announced their presence in world strength sports with a vengeance. The great Nina Romaschkova slammed the discus nearly 170 feet to completely eclipse all previous women's records and many of the early men's records. From then on, the marks collapsed with devastating rapidity with the appearance of marvelous athletes such as Tamara Press of Russia, Lia Manoliu of Rumania, and Faina Melnik, again of Russia.

For the future? It is no exaggeration to suppose that with the nearly religious dedication of the Eastern-bloc nations, the presence of more sophisticated training methods, and increasing awareness of the benefits of modern weight training, the records will soar as far as the weighty platters themselves.

Myron himself would have been pleased.

"Multiple" Mac Wilkins

"I'm not that strong. Compared to most other throwers in my range, my upper body is weak."

I believe that. I believe that about as much as I believe that Lou Ferrigno will deliver the John Locke Lectures at Oxford next year. How does anybody that "weak" toss a four-pound, eight-ounce plate of recalcitrant wood and iron over 227 feet? For that is exactly what Mac Wilkins, the "enfante terrible" of American track and field, has done re-

Mac Wilkins, one of the best discus throwers around. (Warren Morgan)

peatedly since 1973. It is true that big Mac admits to a bench press of "only" 425 pounds and a snatch of barely 285 pounds. On the other hand, Mac normally toys with over 500 pounds in the squat. And somehow, I've not been able to discover his limits in many of the other weightlifting moves (maybe he's not saying?).

The fact is that Mac Wilkins, holder of the gold medal in the 1976 Montreal Games, is probably the strongest, most versatile and vocal athlete in the history of track and field. His career did not really begin until around 1973, when, at the University of Oregon, he graduated as a college hero, having made a name in tossing the discus, shotput, javelin, and hammer. "Multiple Mac" they called him—the man who won the NCAA discus title, finished third in the shotputting, and garnered the AAU discus laurels.

Yet a lot of guts, determination, injury, and failure marked the early years. Every weekend at the University of Oregon, Mac and coach Bill Bowerman faced each other in javelin competition. "By the time spring came around I was throwing around 240 feet," says Mac. He started the next year in college with a throw of 255 feet—impressive but still quite a ways from world-class veterans such as Nemeth and Wolferman. But around that time, Mac incurred an injury that permanently changed the record books in track and field. The intensity and determination that Mac applied to all the throwing events finally caught up to him. The human body simply cannot tolerate the mutually contradictory explosive efforts needed in the shotput, discus, and javelin. In the second meet of his sophomore year, big Mac split his deltoid muscles, thereby putting the quietus on his javelin career.

Happily, everything has its opposite; for every evil there is some good. The elimination of the javelin allowed full rein for his powers, still latent, in the discus. It was that injury, in fact, that sparked the drive to the NCAA and AAU titles mentioned above. But, though his discus career surged powerfully, the 6-foot, 4-inch giant realized it was still a long way to Montreal and the Olympic crown. As Mac describes it, his technique often let him down. "I've always been competitve, but I wasn't really able to show it because of technical aspects. I feel that technique was the only thing that kept me from throwing a world record in '75 or '74." (Still, one wonders: American javelin star Kate Schmidt complains about her technique also. Dare we think the throwing events are mostly a matter of mighty muscles?)

Unpredictably, Wilkins' meteoric career slowed perceptibly after graduation. He entered graduate school at Oregon and did manage to finish second in the discus at the AAU meet in 1974, following this with another unspectacular second-place finish the next year. Not accustomed to finishing second, Mac finally journeyed to Europe to try his hand with the best weight-throwers of the continent. However, his attack on the world records proved to be somewhat less than inspiring. He participated in Finland and emerged a dismal fifth. Then came another pivotal point. Finnish discus star Markku Tuokku took Mac under his wing, training, eating, and practically living with him nearly a month. Results did not wait. Mac journeyed to Sweden and promptly demolished all competition. Ricky Bruch of Sweden befriended him, inviting Mac to stay with him at his home. With amenities like scientifically-styled training rooms and mountains of food, Mac was in his element. With Bruch's encouragement, Mac threw a personal best of over 219 feet by the summer's end. Then came his greatest year, 1976.

That spring he set a world record of 226 feet, 11 inches at the Mount SAC Relays in Walnut, California, followed shortly thereafter by three consecutive detonations of his own world marks, rocketing the platter 229 feet, 230 feet, 5 inches, and finally 232 feet, 6 inches. Mac's 1976 season throwing average was over 218 feet, while his top ten throws averaged over 228 feet. Though arch-foe Schmidt of East Germany had scored a cataclysmic 233-plus feet by 1978, that '76 season has to rank as one of the stellar accomplishments in the discus. Just for dessert, Mac promptly opened the next year with a shotput of over 69 feet, a personal best at that point, and proof of his world-class abilities in that event as well.

Though the Olympic year opened extremely well, it was nearly ruined when Mac injured his clavicle at the Los Angeles Indoor meet. For several weeks, weight training and competition were impossible. But solutions came quickly. Like many, Mac turned to chiropractor Dr. Leroy Perry. Though universally shunned by the clannish and paranoid medical profession, chiropractors of great distinction, such as Perry and Dr. Franco Columbu of California, are eagerly sought after by athletes for their uncanny ability to strengthen muscles within seconds (as Dr. Columbu so convincingly proved to this author!). According to Mac, "If it wasn't for Dr. Leroy Perry, I don't think my body could have responded to the stress I put on it. . ."

After Dr. Perry's "magic," the rest was smooth sailing for the greatest of the age. He headed for the weight room with aplomb, on the highly confirmed assumption that overwhelming strength was a *sine qua non* for success in the discus. Mac has studied the role of strength in the field events so thoroughly that he has developed his own philosophy regarding it. He distinguishes two sorts of power: the one imparted by skillful chiropractic therapy, and the other the strength involved in throwing or lifting heavy weights. Regarding chiropractic power, Mac testifies that ". . . it's really a hard feeling to describe; there's just a lot of energy flowing through you. Dr. Perry uses a number of techniques which anyone can do, but what makes them unique is the results they produce. That's the kind of person he is." For the other kind of strength, it's hours and hours in the weight room. No dynamic tension or disco-body shapers for Mac—just tons of heavy iron. He will lift an average of five times a week during the heavy season. Fundamentally, the Wilkins routine consists of heavy squats with, as mentioned earlier, weights of over 500 pounds. He then does sets of power cleans, regular cleans, snatches, and—for that all-important upper-body power—endless sets of bench presses. As with most strength athletes, he varies the routine throughout the year. Suggestive of the great Olympic and powerlifting routines, Mac gradually decreases the repetitions and increases the poundages until he's at the point where he does extremely heavy weights for as few as two or three repetitions. It is these painful and excruciating "reps" that built the ultimate power Mac brings to competition.

An important facet of such weight-trained strength is that the muscle so built declines only very slowly. Powerlifters and bodybuilders have been known to go for weeks and months with little or no appreciable reduction in size or strength. Franco Columbu, the distinguished Italian bodybuilder and powerlifter, will lay off training for weeks while on the road promoting books. The Soviet colossus Vasili Alexeev often utilized nothing but freehand calisthenics during his training sessions in York, Pa. prior to a recent "Record Makers" competition in Las Vegas. Though big Mac doesn't usually go for that long without caressing a barbell, he has sometimes gone for over a week. As he explains, "It's just something to maintain . . . But you have to be careful, since you can only maintain for so long." He adds, though, that training for the weight-throwing events is to some degree a matter of finding the proper balance. Overtraining is possible. And Mac did precisely that, with nearly disastrous results, for the AAU championships in 1976. As he put it, "I was overtrained. I had trained right through the meet. I could easily have broken the world record but for the very heavy lifting I had been doing right up to the competition. . . ."

Another dimension of the weight-thrower's program is diet, as it is for any superathlete. Surprisingly, perhaps, a 250-pound muscular marvel is made, not born. So it was with Mac Wilkins. In his University of Oregon days he was a relatively "spindly" 210-pounder. The tremendous size added since is, to a large extent, the consequence of a somewhat unappetizing protein-laden fare of tomatoes, soybean extract, and whole mackerel—an inspiration of his college roommate, Craig Brigham. Naturally, Mac's palate is not always in the mood for training dishes. At his famed "Two Big Guys Mountain Games" track meet in San Jose, he is rumored to have downed considerable quantities of megacarbohydrates in the guise of enchiladas, garlic bread, and spaghetti.

The final link in the winning chain is a comprehensive study of technique and mental attitude. His world-shattering ability is a result of powerful muscular development, speed and finely-honed grace, coupled with an overwhelming will to win. As he comments, "My main assets are mental strength, leg strength and explosiveness. My reactions are good . . ." Important though muscles may be, size *per se* is not a crucial factor, unlike Olympic weightlifting, powerlifting, or Sumo wrestling. "The day of the big, fat throwers is over . . . At my level, technique is just about everything. It's really not as difficult to gain strength as it is to throw for distance. You must, however, develop both the technique and the mental power to use it. That's about the whole ball game." And Mac attributes his lifetime best toss of 219-plus feet at one stage in his career to the positive mental attitude developed by his association with Bruch. "He and his surroundings were inspiring. They provided me with a different attitude—more motivated, more positive. I had had my share of self-doubts." As he explains, however, correct mental attitude is not something that erupts from nowhere minutes before a meet. "Getting up for a meet takes weeks. Every competition I am in is the accumulation of what I have done before."

On a rather different note, there is the difficult and painful issue of his notorious antics at the Olympics, culminating in his famous remark, "I would like to see East Germans win all the medals. Maybe that would shake up our people a bit." A potential "Commie" sympathizer? Political turncoat? No. As Mac later explained, the remarks had nothing what-soever to do with ideology or politics. The trouble lay in a perceived indifference and incompetence in the very fiber of United States amateur athletics. His feelings on the matter began to surface just before his event in Montreal. With barely a week remaining, Wilkins and fellow shotputter Al Feuerbach refused to accompany the rest of the team from the training headquarters in Plattsburgh, NY to the Olympic site. Doubtless, Feuerbach's bad experiences had influenced Mac. In Munich, 1972, the former's relatively poor performance (fifth place) had persuaded him that the Olympic site was not the place to train. Although Mac and Al temporarily followed the whims of the Olympic hierarchy by checking in at Montreal, they soon realized that the ruling powers had no real authority to keep them from training where they wished. The duo promptly set out for Three Rivers, a beautiful and inspirational location in Quebec that geared and girded them magically for the coming battle.

For a long time both Wilkins and Feuerbach avoided all reporters, keeping the lid on their expeditions. But when Wilkins grabbed the gold and Al took fourth in shot, the former let loose. "I'm very embarrassed to associate with officials like that," he stated, flinging venom at the United States Olympic Committee. And when asked if he was proud of having garnered the gold medal for the United States, he again retorted, "No, I'm proud that I won it for Mac Wilkins." That, in short order, was followed by the famous quote about the East Germans winning the medals. Then all hell broke loose. Suddenly young Hercules was a "sinister American" and a consummate and hardened traitor—getting free support from the United States while simultaneously and irresponsibly knocking those whose generosity had made it possible. That, of course, was nonsense. As Mac explains and as is evident from the record, he earned his prow-ess every step of the way from Oregon to Montreal. Indeed, the whim-sical bureaucratic tactics regarding Mac's training site had nearly cost him and the good ol' USA the medal! The reference to the East Germans was simply a reference to the adequacy of their training systems. As Mac had to explain *ad nauseam*, "Of course, my intention was never political in any way. The East Germans are a Communist nation, per-haps the harshest form in existence today. I have my choice of where I live and I am living in the United States. My statement was not an ap-proval of their political system. As a matter of fact, it was not even a complete approval of their sports system, where they select and plug their athletes into specific events. Their athletes don't have much choice . . . The East Germans have a dedicated and efficient sports sys-tem. I don't like inefficiency, especially when it involves my perfor-mance. I approach my event with a lot of integrity and I feel that too many of our officials in Montreal did not care . . . " According to Mac, the non-concern of the sports organizations in the United States runs deep, penetrating to the schools themselves.

Among other things, Mac points to the same problem many athletes and college coaches had mentioned while being interviewed for this book. From their comments, I came away with the impression that it may be *the* number-one problem in United States athletics, the overem-phasis on team sports. "Too many efforts are directed towards team sports. There, you either end up as a highly-paid professional or you become a spectator," explains Mac. Agreed. This writer, for one, has long argued that team sports, by contrast with individual sports, actually erode the concept of sportsmanship and individuality that is often touted as central to the American way of life. That, of course, is a topic in itself. Yet the canons of fair play are certainly being violated when one as great as Mac Wilkins is maligned because he happens to have a concept of sport that runs contrary to the American public's bizarre reverence for the likes of baseball and football. For Mac Wilkins, sports means individuality.

Still the critics howl. When Wilkins embraced the East German superstar Wolfgang Schmidt (who took the silver medal), the same accusations began to fly. In truth, however, Wilkins and Powell (also of the United States) had had a long-running dispute, which had many times broken into open warfare. Schmidt, on the other hand, was simply a good friend of Wilkins and they were known to have socialized on numerous occasions. Again, it was the team concept Mac feels is so damaging to American athletes. As he later explained again, "Well, what can you say about the hypocrites who say you are not supposed to be nationalistic in the Olympics, then criticize me for putting down a teammate? Sure, Americans are team-oriented—and a conflict results. But the discus is, after all, an individual event."

Eventually, Mac triumphed both on and off the field. The balance of the moral universe was righted when, in 1977, the President's Commission on Olympic Sports issued a proclamation attacking the structure of amateur athletics in the United States for precisely the faults Mac had pointed out. His intense interest in social and political wrongs extends even beyond the domain of competitive athletics. When he spoke to the legislature in his home state of Oregon, he said nothing about sports but, instead, roundly thanked them for leading the country in conservation legislation.

Maybe friend Feuerbach summed Mac up best. "When Mac gets upset about something, he'll yell about it." Isn't that what makes an American, after all is said? We're probably going to hear a lot more from the genial giant with the guts to stand the athletic cosmos on its self-satisfied nose.

One wonders: How far could Mac throw a discus from the Oval office?

Women Discus Hopefuls

What else can be said about 26-year-old Evelin Jahl, Olympic winner in the discus, holder of the world's record with a toss of 70.72 meters (232 feet) in Dresden in 1978, the gold medal in Moscow, and the solidly entrenched "femme fatale" from East Germany? Lots. For one thing, she is being furiously pursued by a number of worthy competitors, all of whom deserve mention in this section.

There is Lynne Winbigler, 27-years-old from the USA, who set a new American record in the weighty platter toss with a heave of over 189 feet

East Germany's Evelin Jahl, Olympic discus champion. (ADN-Zentralbild)

recently. Though still far behind the great Jahl and others, Winbigler is learning fast. Unlike many of her colleagues in weight events, Winbigler has had good, in fact fantastic, coaching in these events. Her mentor of recent vintage is none other than the heroic Mac Wilkins himself. Before studying with Wilkins, her discus was cruising a lazy 130 feet. Soon after, however, she was reaching an impressive 174 feet! Still, to reach the heights of Melnik, Veleva, and Jahl, the entire American philosophy of coaching is going to have to be revised.

As guru Wilkins explains, "Training programs prescribed by so-called qualified coaches have been too easy on women. The people who run our amateur organizations have an outdated view of what it takes to achieve maximum potential. They are very petty, jealous people trying to maintain the status quo, which means their power, but they are not interested in promoting efficient athletic achievement." (Wilkins speaks of the entire athletic program and not just the women's part.) Wilkins probably has a good point. For whatever the case for athletes generally, his efforts with Winbigler have manifestly paid off. He does not take credit for himself, either, being quick to point out the great gains Winbigler made under the renowned West German coach Christian Gehrmann, trainer of woman shotputter Maren Seidler.

Nor is Winbigler the only non-Eastern European hope. There is lithe Lisa Vogelsang, a slip of a thing compared to some of the European amazons, but with enough strength in her pinky to crush the life out of three average men. Starting in 1976, she was dubbed a "natural thrower," a characterization that she readily endorses. Progress was phenomenal and within seven months she was exceeding all American records, finally taking a seventh place at the Olympic trials in 1976. Yet the coaching problems remain. She, like veteran Kate Schmidt in the javelin, has a technique that would not carry an ordinary athlete beyond the championship of Smoky Creek Corners, New York. Unfortunately, she, unlike Schmidt, has not been able to pierce the very highest rungs in her sport, despite unlimited talent. "I saw a film of one of my throws that was over 185 feet, and the technique was absolutely horrid," she admonishes herself. But the plucky American persists, and it is very likely that with any kind of good coaching in the future both she and Winbigler will be increasingly dangerous opponents.

These up-and-comers just might make Jahl want to take up chess.

8

Javelin: The Most Unpredictable Sport

The muscles tense, sinewy power exudes from every pore. Arm and body poised, the run is short. A final mighty heave and the deadly shaft is hurtling to its destination. A primitive aborigine stalking wild mammoths? No, the scene is the contemporary replaying of an ancient art—the strength event that may, more than any other, be tied most closely with early, arboreal man. That being the case, it is little wonder that the Finns and Swedes have traditionally dominated the javelin throw. The spear was important in their daily life, subsisting as they did in the forests, among the lakes and streams. Of necessity they developed inordinate skill in throwing the deadly weapon.

And since the javelin, along with the discus, was one of the two throwing events included in the ancient Greek pentathlon (going back to the 5th century BC), it is of enormous historical interest. Unlike the discus, however, we have no "Discobulus" depicting the form of the javelin-thrower. There are, however, carvings on various artifacts, as well as two small statuettes in the Metropolitan Museum of Art. Among these is a statue, "Doryphorus" (spear-bearer), fashioned by Polycleitus. Although the original "Doryphorus" has been destroyed, there are at least seven marble statues, 17 torsos, and 36 heads extant. Though it is regrettable that we, unlike historians of the discus, have no single model for the javelin, it's possible to gain some insight from these various embodiments as to what the ancients accepted as "correct" form for the sport. And like the "Discobulus," a glance at the "Doryphorus" reveals something of the aesthetically ideal athlete, squaring to a large degree with Plato's descriptions in The Republic and other works.

The javelin's roots can be traced even farther back in recorded history. Like the shotput, there are reports of javelin-tossing as a "peaceful" pastime in the Tailtin Games in Ireland in 1829 BC. It was known to have been practiced strictly as an athletic event by the Scandinavians in the

Tom Jadwin was UCLA's top javelin thrower in 1979.(Department of Intercollegiate Athletics, UCLA)

old Scandinavian Northern Games of the Vikings. Most likely it was Per Henrik Ling who revived the sport in Scandinavia (around the same time Scandinavia gave gymnastics to the world). Naturally, the number of occasions on which javelin-throwing was practiced as an athletic, peacetime event does not negate the fact that the "peacetime" uses came directly from wartime applications. The javelin was, and indeed is, in some parts of the world a feared combat weapon. So much was this true that the earliest practitioners of the sport tried, unlike in the shotput, for instance, for accuracy as well as for distance. So much technology went into the mechanics of accuracy that a device called an *amentum*, functioning like the tail of a kite, was added to improve flight stability.

The development of the javelin follows a pattern similar to that of many other strength sports—early domination by one nation or group of nations, with gradual diffusion of virtuosity and technique into many lands. (A classic modern example would be the Soviet's gradually catching and finally surpassing the Americans in Olympic weightlifting.) Surprisingly, however, the Swedes and Finns have doggedly maintained their historic presence, if not their early domination in the sport. A recent issue of *Track & Field News* puts a number of Finns among the very best.

Among the earliest important events of modern times was the Swedish championships, held at Halsingborg in 1896, the great Andersson being the winner with a toss of 202 feet (rather uninspiring by comparison with today's tosses of over 300 feet!). Thus, the javelin-throw must be reckoned a comparative newcomer to modern track and field competition. It did not appear in the modern Olympic games until 1906, and the national AAU championships did not feature it until 1909. In fact, the first real superstar in the sport did not come along until 1915, when Jonni Myyra of Finland dominated the event in the Olympiads of 1920 and 1924, smashing several world's records. In 1920 he reached his zenith, uncorking a toss of 219 feet, 1½ inches.

The traditional Swedish-Finn struggle for supremacy manifested itself strongly in 1924, as Myyra lost the world's record to Gunner Lindstrom of Sweden. Yet the pendulum was not to be stilled: Pentiila, the Finnish giant, brought the world's mark back to his native soil in 1927 with a toss of 229 feet, 3 1/8 inches. Yet the record bounced back and forth between the two great peoples with blurring rapidity. Lundquist of Sweden captured the Olympic crown the very next year, only to lose it in the subsequent Olympiad when Matti Jarvinen catapulted the spear into another dimension with a toss of nearly 239 feet.

There was essentially no American to round out the scoreboards in these early days. A good deal of nonsense has been written about this bizarre absence—references to "hereditary advantages" of the Scandinavians, etc. Without further dignifying such absurdities, one could suggest that the most likely explanation has to do with throwing styles. Older greats such as Myyra simply had better technique than the Americans. Ralph Rose, for example, though he was a colossus of the human form and a record-holder in the shotput, won the first AAU national javelin toss with a throw of only 141 feet, 8 inches. Rose, like so many Americans of the day, did not precede the throw with a run, thereby losing all the advantages of that method. Even when the Americans began to participate in greater numbers, the reluctance to use the Finnish running style (which is today essentially the only style seen) and the momentum it imparted slowed their progress considerably. Jim Demers of the University of Oregon, though he threw a credible 220-plus feet in the early '30s, was nonetheless far behind the best Europeans.

Yet the latent power of the Americans eventually surfaced. In the 1932 Olympiad Lee Bartlett gave fair warning to the Europeans by finishing fifth. From then on, the inexorable march of the Americans continued, reaching its climax in the 1952 games when the legendary Cy Young of the United States blasted the previous Olympic mark, hurtling the shaft a staggering 242 feet. And although the traditional Scandinavian virtuosity flared in the 1956 Olympiad when Danielson of Norway smashed the record with a throw of over 281 feet, there was a gathering force building in the East that was to make itself felt throughout the next decade: Viktor Tsibulenko of the Soviet Union captured Olympic laurels in 1960, throwing the javelin nearly 278 feet. And in 1968, Yanis Lusis, also of the USSR, reached a distance of nearly 296 feet. Very soon after that, the virtues of the Eastern European system staggered the Western bloc when a woman, the great Ruth Fuchs of East Germany, flung the spear easily over 200 feet in both the 1972 and 1976 Olympiads.

At this point, some comments are appropriate about the mechanics of the javelin. Most observers recognize intuitively that tossing it differs in character considerably from throwing the shotput or discus. For one thing, the nature of the javelin is such that aerodynamic considerations are crucial. It is widely acknowledged to be the most erratic and unpredictable of throwing sports. Certainly the lightness of the javelin and the extreme degree to which air velocity affects it is a factor. In fact, it is difficult to find accurate accounts of javelin-throwing records from ancient times. An odd passage by Statius tells us that the length of a chariot run was "three times a bow shot and four times a javelin throw"—not exactly the epitome of precision. Even today, with all of our standardization and scientific designs, *Track & Field News* still writes of the event as follows: "One problem with trying to forecast the javelin with any accuracy is the simple fact that it may well be *the* most inconsistent event in track. Because of the flighty nature of the event, one unexpected throw by any athlete can upset all the most thoroughly researched charts." In my judgment, this is correct. Consider, for example, Miklos Nemeth of Hungary—a virtuoso of the spear if ever there was one. In Montreal he hurled the javelin nearly 309 feet, yet has not come within twenty feet of that throw since then! The odds of any similar phenomenon occurring in the shotput are astronomical, to say the least.

Yet, because of the erratic nature of the sport, the role of muscle-power and technique has been a matter of some confusion and controversy. In earlier days, the sport was certainly regarded as a muscle sport *par excellence*. The great Ralph Rose, national shotput champion, threw the spear by sheer brawn alone in the first national championship, being unaware of the role of technique. Yet he won. Soon afterwards, however, many argued that technique was at least as crucial as strength. Indeed, there were instances when a smaller man, thoroughly conversant with European technique, bested a larger man. Even today, javelin-throwers tend to be less bulky than, for example, shotputters, as a comparison of monstrous Al Feuerbach with the relatively slighter Miklos Nemeth shows.

Perhaps the best answer to the strength-technique conundrum rests on noticing that strength seems to play a different sort of role in the javelin than in other strength sports. The javelin does not require the short-term "explosive" strength that the shotput requires. Rather, strength is built through a long run and converted into momentum, which in turn is transmitted into the toss. It's not, as one sometimes hears, that the javelin requires speed rather than strength, but that the powerful run of a world-class thrower amounts to a certain kind of might in itself. Strength and speed are really two sides of the same coin. There are any number of sprinters capable of squatting with weights in excess of 400 pounds. The great skier Karl Schranz is known to be capable of full squats with 400 pounds though skiers and runners are thought of more as "speed" athletes than "strength" athletes. I don't want to suggest, of course, that the javelin requires only great leg strength. Upper-body power is so important that a few world-class throwers seem to have gotten there with that alone. Kathy Schmidt is a good example; though relatively short on technique, she manages to stay in savage pursuit of arch-rival Ruth Fuchs, of the German Democratic Republic, on upper-body power alone. As champion shotputter Oldfield said of her recently, ". . . all arm. If she gets some speed, she'll hit 240 feet." Not surprisingly, Schmidt is an advocate and practitioner of systematic weight training.

Ultimately, the proof is in the pumping: It is widely known that the very best in the world all go through unending sets of bench presses, squats, heavy curls, snatches, etc., to develop the overall strength necessary to squeeze out that extra micrometer. And nobody is better at squeezing out those distances that the great Curt Ransford and the eternal duo of Fuchs and Schmidt.

Curt Ransford

It takes a certain kind of *chutzpah* to throw the javelin: Wind, temperature, and lady luck all conspire to turn the event into a crap shoot. One day you're on top of the world, next day you're a has-been. According to ace thrower Kreider of Army, you can lose 20 or 30 feet from the smallest technical error. You've got to be pretty sure of yourself to go out in front of thousands, knowing abject humiliation waits at your elbow. Enter Curt Ransford. He's got it, that all-important technical mastery and self-assurance that stamp the powerful athlete as the most promising young javeliner in the world.

Yet to be defeated in a dual meet during his tenure at San Jose State University, he was the lone returning NCAA point scorer for San Jose State in 1979. He finished sixth in the nationals that year, compiled an amazing 245 feet, 6 inches versus Arizona and thus gave the world the

Curt Ransford hurls the javelin with the best.
(Athletic Public Relations, San Jose State University)

14th longest throw by any collegian. A transfer from Spokane Community College, he grabbed the Washington State JC title in 1978 under coach John Buck, building solidly on football and track days as a preppie at Moses Lake High in Washington.

Most recently, the peculiar art of the spear worked to Curt's advantage in spectacular style. Though he'd managed a qualifying throw of only 243 feet, 4 inches, 22-year-old Ransford uncorked a staggering toss of 269 feet, 3 inches for a career best and the NCAA title on June 7, 1980. At that rate, it looks very much like he's a sure bet for Los Angeles, 1984. On his victory, Ransford commented, "I wanted a personal record this year. And I wanted it in this meet." Did I say "sure bet" for the Olympics? Can't forget Dame Fortune and the whimsies of the shaft. But one thing is for certain: The heavily-muscled student from San Jose has as good a chance as anybody in the world for the gold at the next Olympiad.

Fuchs & Schmidt—A Dynamic Duo

She's in the winner's circle not for herself, you understand. That would be "antidialectical" and "decadent." Though nobody has bothered to tell poor Ruth Fuchs that such Marxist jargon has as much philosophical substance as an Agnew campaign promise, her coaches have told her plenty about the javelin. And that savvy has enabled her to emerge victorious time and again, for "the people who pay the taxes that enable me to compete." Vintage East German rhetoric: the 5-foot, 6½-inch amazon speaks easily and hypnotically of the virtues of her way of life. For years a member of the Communist Party and honorary delegate to the Ninth German Democratic Republic Party Congress, Fuchs, along with the Olympic gold medalist Colon of Cuba and Biryulina of Russia, is one of the world's best woman's javelin throwers and an absolutely uncompromising supporter of the system that spawned her. Fuch's ego has yielded to politics—machismo absorbed by materialism. As the blond-haired superwoman describes it, the East German system allows total freedom "to go to the stadium and train without putting a single *pfennig* on the table."

It must be conceded that the system works. There's no way most Americans can keep up with them: Maren Seidler is a classic example. The darling of the shotput has lagged behind the great Ivanka Christova of Bulgaria, not for lack of talent but due to sheer lack of support. A virtuoso of technique, the Paganini of the spear, Fuchs spends every moment of her waking life in pursuit of that extra half-inch that will guarantee victory. For the past few years she has spent most of her time in a training camp run by the govenment and is coached by the best East Germany has to offer. To date she boasts of being the oldest gold-medal winner of the European championships of 1980, having set a new European record with a toss of 69.16 meters (227 feet), missing the then world mark by a bare 16 centimeters (6¼ inches). (The record now is 229 feet, 11 inches.)

Of course, not all is smooth going for the imperturbable German athlete. Just behind her is the great Kathy Schmidt of the United States, who chain-smokes and quaffs beer while most of her opposition shoots up on anabolic drugs. With her erratic training methods, sloppy diet, slow approach, and poor technique, she still manages to pose a formidable threat to Fuchs. In 1972 she propelled the spear an astounding 200 feet, 6 inches to set an American record. Recalling Oldfield's prophetic remarks, she is indeed "all arm." If she can couple some velocity with that jackhammer arm of hers, she may well hit the 240-foot mark. The 240-foot goal remains in the future, however, though Schmidt has in recent times shocked the doomsayers and the pundits. Those who once advocated changing her nickname from "Kate the Great" to "Kate the Mediocre" and pointed sneeringly at her disappointing third-place finish in Montreal and the unspectacular early-1977 season are now eating their lines. On September 11, 1977 at Furth, West Germany, all the jokes stopped. She was confident that day; everything was working right. The "kharma" was with her and she was tuned to a state of physical and psychological perfection. Nothing could go wrong and nothing

Gold medal winner Ruth Fuchs of East Germany. (ADN-Zentralbild)

did. Adopting her usual approach, she put everything she had into the heave and damn near rocketed that javelin into Ruth Fuchs' backyard. Sixty-nine meters—a new world's record. As Schmidt described it, "For the last ten years—my whole career—I have thought about what it would feel like to break the world record. I was so overwhelmed, having thought about it for ten years. All those thoughts fell on me at once and I just couldn't handle it." That world record shattering throw of 227 feet, 5 inches had bested arch-foe Fuchs' mark by a solid eight inches. It left her as the only non-Eastern European to hold a world mark in any of the field events for women. The cigarettes and beer had apparently paid off.

Still, Fuchs and others, the aforementioned Colon and Biryulina, remain unfathomably powerful opponents. Who will emerge finally as the superhero of the century—the hard-training, dour Communists or the inordinately talented superwoman who trains on tobacco and guts? Nobody knows, but one thing's for sure: Somebody is going to have to rewrite the training manuals.

9

The Hammers of Thor

According to myth, Achmat, the last of the great Turkish emperors, made a toss of the hammer so awesome that two enormous pillars were set up in Stamboul—one to mark the origin of the toss and the other to mark the landing. The myth is typically fanciful and indicative of the reverence that great feats of strength have spawned throughout history. There is little doubt, though, that Achmat's feat bears virtually no resemblance to truth, but there is also little doubt that tales of Herculean power have and will continue to enchant generations.

Hammer-throwing, the next of the great quadrivium of strength sports surviving into modern track and field, is at least as old as discus-throwing a "sledge" (an early version of the hammer). Throwing the great scholars. There are drawings showing King Henry VIII of England throwing a "sledge" (an early version of the hammer.) Throwing the hammer developed into an important and highly competitive event in the Scottish Highland Games, the competitions that figure so importantly in the history of muscle. In the early days of the Highland Games, both a lighter and heavier hammer were used, the heavier approximating a weight of 25 pounds. Probably the greatest practitioner of the hammer toss in the Games was Johnstone, who managed a heave of over 84 feet in the late 19th century with the 24-pound hammer.

According to David Webster in his *Scottish Highland Games*, the hammer toss, like so many Scottish, Basque, and other strength sports, originated in pragmatic problems early Scotsmen faced in the agrarian Scottish countryside. As Webster explains, "One piece of athletic history known to all who have even the slightest interest in Scottish Highland Games is that throwing-the-hammer competitions were actually derived from throwing a proper long-shafted hammer intended strictly for utilitarian purposes and not for sport."

Oddly, the history of a hammer, more than any other field sport, seems to be a comic study of the bizarre and the quixotic. There is a recorded incident where the great—possibly *the* greatest of Scottish champions—Donald Dinnie tossed the hammer with such incredible

The modern "hammer" is an iron ball on a handle of wire. (Paul Lyman)

force and speed that it sailed dangerously over the heads of a surprised crowd, finally coming to land at a safe point beyond. (On another day, the great Dinnie managed to toss the hammer, or "sledge," directly into the unwilling corpus of a photographer from *Health and Strength* magazine, who luckily escaped unscathed.) Because of such incidents, the old style, where the hammer-thrower turned his body several times to gain momentum, was eventually eliminated. In its stead the hammer was, for a long time, thrown by the "pendulum" swing. With this method, the athlete swung the hammer in a partial arc in front of the body, much like an actual clock pendulum, before releasing it at the top of the swing. Though it protected the crowds because of the greater control it allowed, the style scarcely led to the record-shattering distances of the older style. For that reason still another modification appeared, where the hammer was swung in a figure-eight pattern, to add to the force and momentum of the "pendulum" release. Today, the swinging of the hammer around the head, similar to the Olympic style, is a well-known sight in Braemar and Aboyne every year.

Still, the great Dinnie made a number of outstanding throws even with the inferior styles and nonstandard implements. At Couper Angus he flung the 16-pound hammer a whopping 132 feet. And at Glasgow Dinnie propelled a 17-pound hammer a full 110 feet, forever establishing himself as *primum inter pares* of the Scottish gamesmen. Dinnie was not, of course, the only athlete worthy of mention. The great Tait made a toss of 136 feet with a 16-pound hammer sometime later on. Though he bested Dinnie's previous mark, some questions have been raised as to whether the feat can compare with Dinnie's, since the latter used a hammer with a shaft so long that a trench had to be dug to allow Dinnie to complete a full arc. A number of contemporary Olympic-style hammer virtuosos feel, however, that this is not as significant a factor as one might think. At least many do not think it sufficient to account for a full four feet of distance, when records are broken today by mere fractions of inches.

Whatever the case, "All these great throws," according to Webster, "fade into insignificance when compared to Bill Anderson's throw at Lochearnhead in 1969. On that occasion the 16-pound hammer record was broken by a full seven feet and he broke the hammer head too!" Anderson is, indeed, a super athlete comparable to many modern strength stars, as shown by recent performances in the "World's Strongest Man" competition on TV. And it is worth noting that though Anderson had it over his historical rivals because of superior knowledge and technique, he displayed his greatness by besting drug-trained competitors in the television competition.

Yet the lure of the hammer was not confined to Scotland, though the Scots deserve the credit for the origins and technical development of the modern Olympic version of the sport. In the latter part of the 19th century, the hammer was seen hurtling over the Gothic spires of Oxford and Cambridge. Henry Leeke, a diabolical figure of a man, won the hammer toss for Cambridge in 1869 with a mighty throw of nearly 104 feet. Not to be outdone, arch-rival Oxford soon took their revenge when the great "Hammer" Brown walloped Cambridge's mightiest with a toss of better than 122 feet. The same amusingly chaotic and whimsical "rules" governed the early days of the sport in England much as they

did in Scotland. The size, shape, and length varied kaleidoscopically—
so much so that the judges were in constant fear of death, not knowing
from which avenging undergraduate hand the hammer might next
emerge. Using the same "turning" style prevalent in the early days of
Scottish throwing, an athlete would rotate as fast as possible, control-
ling the flight of the hammer only by the dictates of his subconscious.

As the century drew to a close, a number of shining lights added
themselves to the ever-growing roster of warriors in the sport. It was not
until the opening days of the 20th century, however, that other nations
began to make their mark in hammer-throwing. Flanagan of Ireland, for
example, smashed the world's records in the hammer seven times
while winning three Olympic championships and seven AAU events in
1907. After Flanagan came the mighty McGrath, who completely dominated the hammer for years, winning seven AAU championships and
slamming the hammer into oblivion with an awesome throw of 179 feet,
a record that finally fell in the 1936 Olympiad. With the Irish foray into the
sport, there followed quickly an infusion of foreign blood. Germany,
Japan, Sweden, Finland, and America soon were welcoming the hammer into their borders, only to toss it out with a vengeance! In 1927,
Norwegian great O. Skoeld placed fifth in Paris with a fine throw of over
148 feet, before finally coming to England itself, humiliating the British
with a thundering victory over Nokes in the AAA (Amateur Athletic As-
socation) event of the same year. So far as America was concerned, the
superstar of those days was the mightly McDonald, winning ten titles
between 1911 and 1933, as well as the Antwerp Olympiad by virtue of his
fine abilities in the 56-pound weight-throw. (The Antwerp games
marked the final appearance of the 56-pound hammer toss as a regular
Olympic event.) Other Americans prominent in this period include Fred
Tootell of Bowdoin College, who garnered the crown of the 1924 Olympiad; Folwartshny, who rocketed the 16-pound hammer nearly 179 feet
in taking the IC4A championship in 1938; and Henry Dreyer, who was
national champion in 1935.

By this time the event had been quite standardized by the efforts of
early Irish-Americans such as Ryan and McDonald. And in 1908 the
rules were officially adopted by the *International Amateur Athletic Federation.* As practiced, the "hammer" is really an iron ball with a handle of
wire that must be under four feet long, with the entire apparatus weighing not more than 16 pounds. The throw begins from a seven-foot circle
and legal throws are confined to a 90-degree sector. Also, the athlete
must wear a glove on the left hand. Because of these standardizations,
which allow for the mechanics of the sport to be more easily studied,
accomplishments in hammer-throwing have soared even more than in
eariler years. In recent times, superstars such as Gyula Zsivotsky have
catapulted the weighty ball a full 241 feet, 11 inches, while some 11 years
later the great Yuri Sedyk managed a heave of over 254 feet in the
Montreal Olympiad, only to stretch the world's record to an amazing
268 feet, 4½ inches in Moscow!

In sizzling pursuit are the astonishingly large number of rapidly improving hammer-throwers on the college scene. For whatever reason,
there seem to be more great and near-great hammer enthusiasts than
there are in any other weight-throwing sport. (Perhaps because there
are no women competitors chasing their records, though the ladies may

Yuri Sedyk heaved the hammer nearly 270 feet! (R. Maximov)

take up the sport eventually.) At Princeton, Dave Pellegrini is certainly one of the better weight-throwers around, and will most likely reach world-class caliber shortly. Most recently he grabbed first place in the 35-pound weight throw in the 59th Annual IC4A Championships held at Princeton, scoring nearly half of the team's total points. With a best toss of nearly 70 feet with the heavier weight, he's given fair notice to the world's best. Another great one is Neilson of Canada, who tossed the 16-pound ball over 236 feet to overwhelm the opposition in the '79 National AAU competitions.

The training for hammer-throwing follows the same general pattern as the other weight-throwing events—plenty of resistance work, careful attention to upper-body power, and an eye for diet. (It is interesting to note that Scottish athletes have paid far less attention to the million-and-one training systems and bizarre high-protein diets that so many weight-trained athletes fanatically follow. Nor has it hurt their relative performances any. Could all the alleged "need" for supplements, 400-grams-per-day protein intake and all the rest of the hoopla, be mere "hype" from the manufacturers?) And devoted training does make some difference in many cases; Dave Pellegrini for example, engineered a near-miraculous transformation in bringing his pitifully underdeveloped 110-pound frame to a status that would rival Schwarzenegger.

Lucky for Achmat, these guys don't take to setting up any more pillars.

Ed Kania

Ed Kania had better make his first million fast. When they get through charging him for the paper and ink for all those record books they've had to rewrite, he'll need every penny. A 1979 graduate of Dartmouth College and a member of the Pacific Coast Club, the 6-foot, 5-inch mass of speed and brawn has been sending theories—and distances—jetting into the Crab Nebula for years. His stardom began back at Cheshire High in New Haven, where he lettered in track and was a member of the phenomenal Housatonic track team from 1972 through 1975.

Ed Kania has been breaking hammer records for years. (Athletic Council of Dartmouth College)

After arriving at Dartmouth, he obliterated all standing marks in the 35-pound weight throw, hammer, and discus before being awarded a $1500 postgraduate scholarship by the NCAA. He was one of six Division I athletes to be so honored. In the 35-pound weight throw, he remained undefeated in two indoor seasons while dominating the field at the National AAU championships, the IC4A championships, and the Indoor Heptagonal events. The only blemish on the remarkable record was a second-place finish to the phenomenal Robert Neilson of the University of Michigan in the NCAA. To date, his best heave with the 35-pound weight came at the National AAU Championships with a mark of 72 feet, 2½ inches— one of the finest throws in the world with that weight.

Kania doesn't always like the weather inside either; when it's pleasant he ambles forth to spin the 35-pound weight into orbit for a new New England record. In 1979 he found time to serve as captain of the indoor and outdoor track team and was named winner of the Archibald Athletic Prize, which honors both athletic ability and cerebral agility. As with the tremendous Maren Seidler of women's shotput and Dr. Franco Columbu, bodybuilding superstar, Kania's brain serves him as well as his brawn. Accepted enthusiastically at the Harvard Business School, he'll be taking his protein at "Elsie's" for the next few years. Meanwhile he and teammate Ken Jansson, also a star performer, will be tossing the iron ferociously.

Yuri Sedyk? He'd better keep working.

10

Wristwrestling: Savages of Strength

Afternoon tea at Fortnum & Mason's it's not. First of all there is the Crazy Cricket Eater. No kidding. This guy believes that dead crickets add important protein (they do) and that eating them helps psyche out his opposition (it does). As the *Argus-Courier* in Petaluma put it, "If his arm didn't get you, his breath probably will." Of course, everything gets ridiculous after a point. So the Crazy Cricket Eater switched to live grasshoppers; not on his own, but at ABC's prompting (good for color). But wristwrestling just ain't that easy: The Crazy Cricket Eater failed to make the finals. Then there's the likes of big Cleve Dean, 466 pounds of "good ole boy"—an insanely powerful Georgia hog farmer whittlin' and bendin' his opponents into submission. He will not, most likely, be entrusted with delicate matters of international negotiation. After watching several hours of behemoths munching insects, massive beer-bellied gargantuas spitting and cursing whilst reveling in the mellifluous melodies of crunching bones, cracking joints, and tearing muscles, one cannot but fail to believe that civilization has indeed reached its *terminus ad quem*.

But that's wristwrestling. Or that's "arm" or "Indian" wrestling, or "arm-turning" (English) or whatever the hell you want to call the noise of all that macabre flesh-grinding emanating from California and Georgia every so often. Still, there's a sort of refined honesty about all those out-of-shape bodies striving to mutilate all those other out-of-shape bodies. Though it lacks the refined elegance of high-level bodybuilding, it also lacks the somewhat contrived quality of women's bodybuilding and the hoky pretentiousness of some competitive powerlifters.

Thankfully, there are a minimal number of evangelical, "praise-the-Lord" types in wristwrestling. (I don't know why so many strength athletes get religion when they get strong; maybe they think they're the living reincarnation of Samson, chosen by the Almighty himself.) Also, you don't find steroid use and abuse as much among the wristwrestlers.

The sounds of crunching bones, cracking joints, and tearing muscles are commonly heard at wristwrestling contests. (Larry Heller)

Some of these guys don't even lift weights! Hell, Big Cleve Dean got all of his power pickin' up hogs by the ears down in Georgia. He only started tossin' the iron recently and, Alexeev beware, God only knows how strong the man is gonna get. Who the blazes ever heard of Big "Mac" Batchelor lifting weights? As far as I know, the only weight training the man did consisted of heaving plastered customers out the front door while he was tending bar. And he may be the best there ever was. Not to disparage the weightlifters, of course; some of the best have turned a wrist or two themselves. The great Jimmy Payne, still one of the feared competitors though he is past 50, was as great a bodybuilder as ever there was.

It's hard, almost impossible, to trace the beginnings of this noble sport. Probably the caveman practiced it, given the natural attraction it exudes for every man wanting to test his physical mettle. (Where is the person who has not tried it once in his or her life?) Abe Lincoln was a devotee of arm-wrestling. Ernest Hemingway, the great novelist, may have been an afficionado as well, as he affectionately describes a lengthy wristwrestling scene in *The Old Man and the Sea*. Another great writer, Jack London, also talks of it in some of his works. Many of the old-time Vaudeville strongmen were, not surprisingly, greatly accomplished in the sport. The "Cincinnati Strongman," Henry Holtgrewe, was thought to have been undefeated at the game. The great Hermann Goerner may well have been the mightiest of the old-timers. Edgar Muller, in his book *Goerner the Mighty*, describes his prowess as follows: "In this contest, the six men were seated at a table on one side, whilst Hermann occupied the opposite side of the table. Starting with the first man, he rapidly downed his arm and carried on along the row of men until he came to the sixth. As has been already stated, he flattened the forearm of the sixth man before one minute had elapsed. At the finish of this contest Hermann called out smilingly, 'Next gentleman, please,' but there were no takers . . ."

Joe Nordquest, the one-legged superman, downed the best in the sport until he fell before the onslaught of gymnast and strongman Franklin D'Amour. For sheer *hubris*, none can match Maxick, who claimed in the late '30s he could hold the wrist of the great Goerner. Given the difference in size as well as reputation, this seems just a bit much. Moving into more recent times, "Mac" Batchelor was surely supreme. Not only did he reign undefeated for over 25 years, but it is said that none of his matches ever went beyond two minutes! Perhaps the closest anybody ever came to providing some opposition for Batchelor was football player Earl Audet, who weighed over 300 pounds. Audet's prowess was formidable and many think he easily would have been the equal of Batchelor had he not been intimidated by his great opponent. Be that as it may, "Mac" overcame Audet in Los Angeles in 1946 to become the first official wristwrestling champion in the United States.

Nevertheless, wristwrestling continued in the same chaotic, disorganized state that had always characterized it and, indeed, seemed fitted to its impromptu challenges. In 1962, however, tradition snapped. The town of Petaluma, California, became the site of an annual competition in the sport. At its first championships, the men's heavyweight crown was grabbed by the great Earl Hagerman, who managed to overwhelm men much larger than himself. It seems, on the other hand, that though the sport is "organized" to the extent that there are governing organizations and regular tourneys, the order is only skin-deep. In the best barroom traditions of arm-crunching, the 1978 Petaluma tournament may shine forever as the quintessential example of insanity and mayhem. Spectators threw beer at a tournament director for unclear reasons, the loser of a match smashed the victor in the face, endless disputes broke out over the amount and distribution of prize money and, not to be left out of the fun, a woman's competitor is alleged to have tried to strangle another woman with her girdle. Another got her arm broken.

Though the Petaluma tournaments are the best known, due primarily to television coverage, there are other organizations vying for influence in the game. One, the *National American Arm Wrestling Association* Championship, drew 150 male and 12 female entries in 1978 to its Scranton, Pennsylvania, competition. It was in that tournament that the legendary Allen Turner of Brockton, Massachusetts, won his eighth national title. Though well past 50 he, like his colleague Jimmy Payne, finds age no problem whatsoever.

Possibly the finest and most active organization in the world is the *World Professional Wristwrestling Association*, founded in 1979 by L.B. Baker, Cleve Dean, and B. Whitfield, all men of outstanding skill and accomplishment in the sport. As well as offering its own national events, it is actively engaged in promoting local competitions throughout the United States. At present there are three world champions in the organization and it is likely to become, without dispute, the dominant organization in the sport.

Despite its vague history and disorganization, the sport does seem to have a relatively standard set of techniques. The most common and most intuitive *offhand* way of engaging one's opponent involves the competitors facing one another across a table, clasping hands in a sort of inverted-handshake grip. Each then tries to force the arm of his opponent to the table. In tournament practice, however, the base of the thumb is grasped instead, to forestall the common technique of letting the wrist go limp to dissipate the force of the opponent's thrust. Though there is some latitude in the other rules, they commonly require a table slightly over a yard high, with contestants being required to maintain one foot on the floor at all times. Because of many untoward occurrences and the less than scholarly behavior common among wrist twisters, the major organizations have found it necessary to construct an ever-lengthening list of disallowable practices (eating crickets has not yet been forbidden). The World Professional Wristwrestling Association deems it a foul whenever a participant uses his face or chin to help push, whenever he or she delays gripping the opponent's hand, or generally conducts himself or herself in a way "inappropriate to the sport." (What is inappropriate in wristwrestling mayhem?) A "foul" is awarded for each such offense and three fouls result in a loss.

I suggested before that wristwrestlers train relatively little. Actually, that is not quite fair; though the artistry of the game is unlikely to rival gymnastics in the near future, there is some skill and technique. A common technique consists of the wrestler "locking" his elbow joint to decrease the angle between his upper arm and forearm, thereby creating passive resistance. It does not require great power to hold this position, and the opponent can quickly run out of steam trying to unlock the arm. There are endless psychological tactics as well. Glaring at the opposition in the manner of a killer shark, delaying and repositioning the hand interminably, screaming pointlessly, etc—these are all ammunition for the virtuosos of the elbow. One of the sport's great legends holds that Maurice Saxe, Count of Saxony, was crushed by a woman using such tactics. Mlle. Gauthier, an actress appearing in the *Comedie Francaise* in the late 1700s, is said to have defeated the powerful Count with ease, despite his vastly greater strength. Some experts contend that her personality was so overwhelming (born of long experience facing hostile 18th-century audiences, no doubt) that the Count was lost before he had a chance to position his grip.

It is not surprising that such tactics play so crucial a role in wristwrestling. For the sport is, in a way, a consummate medieval power drama—the ultimate battle between Good and Evil, Ahram vs. Orzmud playing out their timeless war in microcosm over a three-foot table. It is, among all strength sports, perhaps *the* paradigmatic battle of Man vs. Man. Though weightlifting plays a role, though technique has its niche, it remains, for all that, an essentially one-on-one battle.

In weightlifting, bodybuilding, or track and field, the character of the competition is radically different. For one, the vanquished takes something home to soothe the bruised ego. If Sultan Rakhmanov loses to Alexeev, hoisting a "mere" 507 pounds to the latter's 550-plus, there is balm for the wound. When all is said and done, the loser has hoisted a poundage worthy of the gods. Though Alexeev lifted more in Montreal, Bonk took home the memory of standing proud and alone holding a quarter of a ton overhead. He knew well that he had accomplished a feat matched by very, very few in history. He had convincingly shown his superiority over all except the Russian. It is thus with most sports—the loser having "what to show" for his loss.

In arm or wristwrestling the case is otherwise. The loser has nothing. Years of training, diet, and preparation can go for naught in that frail fraction of a microsecond separating victory from defeat. Despite his overwhelming strength, the vanquished can be made to appear as the lowliest tyro. He has no memory of supporting an unbelievable weight overhead, no memory of hurtling a discus farther than anyone else except the actual winner, no recollection of throwing 11 strikes to the winner's 12, no soothing thoughts of tying the record in the mile where the winner breaks it. He has, in short, only the memory of a defeated arm lying alone, humiliated, beneath the conqueror's. In nearly all sports, then, the loser has a compensation prize in the memory and knowledge of what he has done, as well as, to be sure, what he has not. In wristwrestling it is otherwise. Man against man, brutal, savage, primitive ego vs. primitive ego. Eye vs. eye and limb vs. limb. Winner take all—including the soul of the conquered.

In this the sport resembles most closely that other colossal war of the ego—chess. Surprising analogy? Perhaps, but a good one. Did not Fischer himself say that his principal satisfaction in playing chess is to "see the opponent break up inside?" As with wristwrestling, the loser has nothing to show. The most careful preparation of the Accelerated Fianchetto variation of the Sicilian Defense can be crushed by ignorance of Shamkovitch's latest analysis in *Shakhtmaty*. All study, calculation, psychological preparation, and technique are for naught in the face of hasty assault. A weakness, backward pawns for example, can bring an avalanche of humiliation upon the loser.

It is precisely thus with wristwrestling. I don't wonder, then, at the tactics of these odd encounters. It is not because the combatants are less "principled" than those in other sports. To believe so is to misread the essential nature of the struggle, and how it can take far more from a man or woman than all other sports combined. It's All or Nothing. Its practitioners will torture themselves on odd contraptions resembling the torture machines at the Tower of London. There are gadgets that stretch, pull, wrench, and bombard the body with every imaginable form of resistance, and all are utilized at some time or other by wristwrestlers. (There is a fearsome device, for example, that acts as a mechanical hand to duplicate the resistance of actual battle. Truth is surely stranger than fiction!) That is why the great Maurice "Moe" Baker is alleged to have declared that his sport alone was for real men, while other strength sports were for "sissies." In 1979 the mighty Ken Taylor of Quebec took the heavyweight class, but not before charging the platform like a raging animal—roaring, smashing the floor in the manner of a titanic Sumo wrestler, heaving resin bags to the floor and, in general, behaving like a man about to defend the universe itself against all the powers of Darkness. This is why the supremely powerful Karyn Jubinville can descend upon her hapless opponent spitting fire and brimstone and exploding in ecstasy at the moment of victory. This is why one competitor can assault another with, apparently, all intention to kill. It is not the prize, not the championship, but the Essential Man that is at stake, the Nietzschean ego waging war with the inferiors. That is wristwrestling and that's why its participants must reign supreme in the world of raw guts.

Cleve Dean

I don't know why I like big Cleve Dean. Maybe it's because he's a pig farmer and I am enchanted with rural agrarian life, having grown up in the farmlands of Endicott, New York, and its environs. Or maybe it's because he doesn't take crap from anybody. Maybe it's just because he's the best there is in the sport.

Anyway, big Cleve began heavyweight wristwrestling on December 5, 1977, at age 24, at Barrington Ford in Columbus, Georgia. He took home the second-place trophy that day—the only one he owns. Within six months he won his first title at the South Georgia wristwrestling championships in Valdosta. Since then the gregarious pig farmer has taken every important national and world title in wristwrestling. On October 1, 1978, he added to his string of victories, winning the WWC world title in the process, by grabbing the scalp of George Hood on that fateful day in Petaluma. He repeated in 1979, whipping no less than the great Danny Stone of Birmingham, Alabama. In 1979 he crushed big Dan Mason for the WPAA arm-wrestling crown in Kansas City, and mauled big Dave McCay for the 1980 Petaluma Crown.

It's difficult to choose among the bizarre episodes that compose the fabric of the colorful superman's career. We could, as one example, talk about the time big Cleve, during the 1979 championships, stopped the whole shebang by bullying his way onto the stage to grab the mike from a rapidly-wilting M.C., thereby nearly ending the 18th "Walter Mitty Sports Spectacular." Why? Because a network "bigwig" decided that no

Cleve Dean, Georgia hog farmer and champion wristwrestler. (Larry Heller)

one could publicize the names of the sponsors. The competitors didn't take kindly to that, since most of them depended upon sponsor support to participate at all. There were almost more mangled hands than usual that day.

Cleve decided that the TV executives were way out of line and he wasn't about to let down his own sponsor, the Miller Brewing Company. As he put it in his imposing Georgia drawl, "Just like a lot of other guys here, I couldn't have afforded the trip out here unless I had a sponsor. I have got a wife and kids to support . . ." What followed was intense and a bit ugly, but Cleve got his way without any ugliness on his part, and without robbing the fans of the drama they'd paid their five bucks to see. The producers, knowing Dean was no one to trifle with, capitulated. "To hell with it," they said. "Let's start the show."

About Dean's training . . . Hell, what training? The Georgia farmboy has no more to train to stay on top than Minnesota Fats needs to practice his massé shots. But they say you can't just walk off the street and be the best in the world. They didn't tell Cleve, though, so that's precisely what he did. There was just no turning back that 466 pounds of raw, arrogant, primordial might. The crowd doesn't always like him—"Break the pig farmer's arm off!" they scream. But the farmer just sits there stoically, grinning and turning his hat backwards, quite bored. He doesn't care about taunts and he doesn't care about his opponents. He just puts their arms on the table and their souls in his pocket. Danny Stone of Alabama once tried to put the legend's wrist down—how he tried! In the 1979 championships, Stone, giving away 230 pounds to Dean, approached the losing table six times before locking wrists and accepting the inevitable. Dean was becoming visibly irritated; a few more such shenanigans and Stone might have found himself hoisted by the ears in the manner of a fat Georgia hog. McCay's 1980 master plan fared no better.

Just to make things worse for the arm-bending community, it's rumoured that Dean's going to keep lifting weights after winning the recent title. The problem is that he, like the awesome Paul Anderson, already has more natural power than most people acquire in a lifetime of steroids and bench presses. He recently placed sixth in the CBS "World's Strongest Man" competition; with a little weightlifting he could be a threat to every major strength title there is. And with that no-nonsense TV bargaining prowess, maybe he should join forces with Mac Wilkins in the oval office. Yeah, I know what I said at the beginning about big Cleve dealing with delicate international negotiations, but who needs "delicate" with the Ayatollah?

Mildred Choplick

"I'm not Mrs. or Ms., just me."

That sums up the great Mildred Choplick, the *femme fatale* of world-class wristwrestling. A perennial figure on the podium of power, the beautiful and charismatic Choplick has been a dominant force in the game for years. Unaffected and unpretentious, she's bent more wrists than any woman in history. Hailing from Iowa, Mildred found time to

Today women are entering wristwrestling competitions. (Larry Heller)

raise two children and create a loving household while reigning as the queen of wrists, defeating such notables as Stella Cornejo (in just 20 seconds!) in the 1977 tourney, as well as the always tough Rebel Chambers, Jane Carpentier, and Charleen Carmen. With the last-round conquest of Carmen, Mildred, already a national champion, added the world crown to her laurels.

Still, life has had its problems for the awesome wrist-mangler. In the 1979 Petaluma championships, a cataclysmic upset occurred when a relative underdog, Pam Carter of Modesta, California, forced down Choplick's arm with surprising ease. It was a holy war of sorts, winner Carter proudly proclaiming, "It just took the willingness to win and good technique. And I got the jump on her. I had trained hard for one year. I wasn't about to lose." Nor was Carter's victory a fluke. On the way to her crown she mowed down Cynthia Baker (daughter of the heroic "Moe" Baker), among others. Were I a betting man, I'd venture that an eternal struggle, suggestive of the epic battles between javeliners Kathy Schmidt and Ruth Fuchs, is brewing. Carter's good, but Mildred doesn't take lightly to losing. We can bet that she'll be training all the harder, and that future battles will be closer than ever.

Still young for world-class wristwrestling, she's only beginning to discover all her enormous potential. Some more weight training, a little of Mike Dayton's *Ch'i* mind-control exercises for better concentration, and we'll be seeing her on the winner's podium. Of course, there are other obstacles besides '79 and '80 champion Carter; for one, there is the haunting specter of champions from other organizations. Enter one Rocky "The Rock" O'Shaughnessy, a genial but enormous and enormously powerful arm-bender from Fitchburg, Massachusetts. She is so imposing a sight that muggers ask her for her autograph. "The Rock" is thankfully in the Eastern-based *American Arm Wrestling Association* (AAA) for the moment. If she ever decides to move to Petaluma, she could be Choplick's biggest threat.

11

Brawn in Britain: The Scottish Highland Games

THE HEPTATHLON GAMES

"You shall have joy, or you shall have power, said God; you shall not have both."

—*Ralph Waldo Emerson*

The great American philosopher was obviously unfamiliar with the joys of the heptathlon, for these legendary Scottish games of strength have been practiced and enjoyed since the 10th century AD. And according to many afficionados, the heptathlon is among the most grueling athletic contests known to modern man. Its origin is uncertain, save that writers as far back as the 16th century have referred to it, giving many exotic explanations for its beginnings. According to David Webster, the games of the heptathlon were devised for sport and relaxation. Other sources, however, claim a more pragmatic genesis, urging that the games originated as a consequence of the great feats of strength needed for survival. Caber-tossing is said to have emerged from the practice of cutting down trees (cabers) and tossing them across streams as a substitute for bridges. Similarly, the heptathlon contest of lofting a 22-pound sheaf of hay over a crossbar with a pitchfork is thought to have originated out of the necessity for developing sufficient strength to handle large bales of hay with ease in agrarian 10th-century Scotland.

Though the strength sports constituting the heptathlon were for many centuries relatively unknown outside of Scotland, they are fast becoming a popular spectator sport in many countries. At Oxford University there is a Heptathlon Society that meets regularly to hold competitions and socialize. In the US, places like Santa Rosa, Orlando, Alexandria, Virginia, and Ligonier, Pennsylvania, are rapidly becoming hotbeds of this Scottish mania for tossing, pushing, lifting, and heaving great weights. Crowds of over 15,000 are becoming common at the regular stagings of the heptathlons in these places.

The heptathlon is a sterling and sometimes dangerous test of both strength and endurance. It consists of seven events, all of which are completed in a single day. They include throwing a 22- or 16-pound hammer for distance; using a pitchfork to loft a 22-pound bale of hay over a crossbar; putting stones (of varying sizes); tossing a caber for accuracy; chucking both a 28-pound and 56-pound square weight for distance; heaving the 56-pound weight over a crossbar for height; and, finally, Cumberland wrestling—a form of wrestling where the object is to dislodge the opponent so that a part of him other than the soles of his feet touches the ground. Every participant must enter all seven events and points are awarded for first through fifth places. Total points are recorded and the overall heptathlon winner is determined by the highest score.

Because of the extreme tests of physical reserves in the heptathlon, many American athletes find themselves unable to resist the challenge. Brian Oldfield, the great Olympic shotputter, is an enthusiastic and accomplished practitioner of the seven events. He has set the American record for the 56-pound weight toss for height at 15 feet, 6 inches (recently broken by powerlifter Bill Kazmaier). A testimony to the inordinate difficulty of the Scottish games is the fact that many American athletes who have achieved distinction in other strength sports do not measure up in the heptathlon. Bill Skinner has tried and failed to win anything in the heptathlon, though he is a former national javelin champion. George Frenn, the national hammer-throw champion, finds the Scottish hammers a bit recalcitrant. According to Frenn, "Scottish pounds are heavier than American pounds, apparently."

Perhaps the American virtuoso of Scottish heptathlons, Fred Vaughan, captured the essence of these events in his recent statement. "I've always thought that the individual sports are the most fulfilling," he said. "This is especially true of track and field and weightlifting. In Scottish athletics, I feel that I am in a way celebrating the history of both sports. Each began in similar circumstances. At least two Olympic field events—the shotput and hammer throw—can be traced directly to Scottish origins."

The training for these events, not surprisingly, is often as difficult as the games themselves. And in many ways it demands more than the training for Olympic weightlifting. For, unlike weightlifting, the Scottish games make the most exacting demands on both strength and endurance, since the games can easily take the entire day under a blazing sun. Typically, a modern heptathlon hopeful will train about 15 hours per week, usually dividing his time between developing strength, endurance, and technique. The latter is especially important; poor form can cost the strongest man the title. For development of extreme strength, the competitors will do hours of heavy cleans, bench presses, dumbbell presses, squats, and heavy curls. Endurance movements such as jogging, swimming, and bicycling are common as well.

Diet is paramount. As with all strength sports, some heptathlon athletes find it essential to include enormous amounts of protein in their diets—some going to as much as 400 grams per day in the form of chicken, steaks, liver, eggs, and fish. Four thousand-calories-a-day diets

Putting stones is an essential part of the Scottish Highland Games. (David P. Webster)

are used as well, along with enormous amounts of vitamin, mineral, and amino acid supplementation. Milkshakes reminiscent of those of Olympic lifters are occasionally whipped up with abandon—dozens of eggs, nut butters, whole containers of protein powders, anything that might send a 56-pound stone one millimeter farther, are enthusiastically tossed into the blender.

Perhaps the best known and most difficult of the heptathlon events is caber-tossing, one reason being that the cabers, reaching a weight of 150 pounds and a length of 20 feet, are far and away the most unwieldy and stubborn obstacles to victory. It is commonplace for a contestant to run away with all other events and then lose the championship in the caber toss. Alternatively, there are those so enthralled by the allure and insurmountable difficulty of the caber that they acquire mastery in that event to the exclusion of all others. (One example is Felix Popitti of Wilmington, Delaware, who, while winning seven consecutive caber-tossing championships, never attained the overall heptathlon title.)

Oddly, victory in the caber toss depends not so much on distance as on accuracy, though enormous strength is required to get an extremely awkward 150-pound tree trunk into the starting position. In competition, the contestant must maneuver the caber into a balanced vertical position against the shoulders. Assistants are often employed to bring the caber to the vertical position and, when the competitor feels he is ready, he signals the assistants to let go. He then begins a 20-yard run, during which the caber must be kept under control at all times. At the end of the alloted running distance, he tosses the caber end over end so that it lands (hopefully) on its thin end, coming to rest with the thick end pointing away from the thrower in the "12 o'clock" position. The proverb notwithstanding, "close" counts in caber-tossing as well as in horseshoes. As if the mere feat of lifting and running with the caber were not enough to whet the appetite of perfection-hungry Scotsmen, the best Highland traditions demand that the winner be not only the strongest and most accurate, but the one who tosses in the best and easiest style as well. According to caber-tossing champion Felix Popitti, "Technique is quite important, because caber-throwing rules are very strictly enforced."

Although weightlifting is a vital ingredient in the training for caber-tossing, as well as for the other heptathlon events, the unique combination of strength, skill, and grace of movement in the former demand much more than weightlifting by way of preparation. Popitti, for example, was already winning Delaware State Olympic weightlifting and powerlifting titles and several national and state physique competitions before he dared look at a caber. Besides the usual gamut of bench presses, squats, dips with weights, and overhead presses that are necessary to develop the sort of explosive power needed in caber-tossing, many also work hours on end on style alone. "Normal" training for all caber-tossing competitors involves the use of cabers of varying sizes for mastering balance and control. Some will use only two cabers in workouts, the larger one weighing about 150 pounds, or slightly over competition size, and another one of perhaps 110 pounds.

One ingredient for success in caber tossing is the recognition of the importance not only of strength, but of balance. According to Webster, "Keeping the caber in balance during the lift is tremendously difficult and throwers often have to take a step or two backwards to maintain control."

Perhaps *the* equation for success in the caber and, in fact, all the Scottish events, is great skill plus solid bonnie craftiness. As Webster puts it, "Another necessity is to be able to judge the best place from which to make the throw or alternatively to select a good landing place for the caber . . . It takes years to learn all these tricks and caber tossers are usually slow to mature." Indeed, the intelligence obviously required by the Scottish Games and displayed by the competitors is suggestive of the enormous mental effort the Soviet lifters began putting into the technique, dynamics and physiology of Olympic lifting back in the sixties.

Obviously, the Scottish competitors have learned a lot from the Soviet strongmen—or could it be the other way around?

Bill Anderson

"Anderson doesn't just break records, he murders them." That's what they say back in bonnie Scotland about their homegrown phenomenon of strength, Bill Anderson. The great one has been making and breaking his own records for so long in his native Highlands, he must be boring himself. He won his first Scottish Games championship in 1959, followed by triumphs again in 1960, '61, '62, and '63. And only in 1963 did he have to share first-place laurels with Arthur Rowe. He won again in 1970, '72, and '73. He's been compared with Donald Dinnie, as well as with A.A. Cameron and George Clark, but the record clearly shows that Anderson is in a class by himself.

Athletics were always second nature to the great Scotsman, renowned for his trophies in hammer-throwing, shotputting, and some respectable performances as a boxer shortly after his discharge from the service. As David Webster describes him, "I remember the first time I saw him in action how impressed I was, particularly with his hammer-throwing . . . I was amazed at the controlled wind-up for his throw and the terrific speed he imparted into the hammer as he released it. He was the second person to throw but he disrupted the competition completely for his effort added seven feet to the existing record and split the hammer head in two."

Astonishingly, Webster finds eerie parallels between Anderson and Dinnie; both came from the northeast, they were born exactly 100 years apart, both went into the building trade, and both came from large families. Both Dinnie and Anderson traveled widely, though Anderson far more than Dinnie. The awesome Scotsman has competed in Kenya, America, Canada, the Bahamas, Sweden, Scotland, Nova Scotia, Australia, and Japan. Most recently he has become known to Americans for his sterling efforts on CBS's "World's Strongest Man" competition, where he took the scalps of many superb athletes.

Perhaps most amazing of all was the way big Bill handled a few martial arts experts. As any strength athlete knows, judo and karate types are quite fond of running off at the mouth about the devastation that would be wrought were one of their numbers to meet a strongman in battle. (A tiresome lot, these Bruce Lee imitations, as such claims are

Champion Bill Anderson tossing the caber. (David P. Webster)

rarely tested. Can one imagine the spectacle of a karate miniature trying to topple the great Alexeev or the superhuman Franco Columbu? Any karate chop would be quite nicely absorbed, especially in Uncle Vasili's abominable abdominals.) Anderson's own contribution to the dissolution of such myths came when he humiliated a Japanese judo expert before 6000 people at Toronto's Maple Leaf Garden in 1964. So much for the legends of martial arts invincibility.

Far more interesting is the thorough thrashing Anderson handed a giant Sumo wrestler. The great Japanese strongmen are among the mightiest, fastest, and cleverest race of human beings ever to walk the earth. Hence, taking their measure is a feat far in excess to mere trifling with karate experts. The occasion was the Scottish Games held at Toshimaen, attended by some 40,000 people, including a number of Japanese princesses and some British royalty. After the games, David Webster managed to arrange for Anderson to meet one of the biggest Sumatori at the famed Sumo stable of Grand Champion Wakanohana, then holder of the fabled Emperor's Cup, the highest accolade in Sumo. Though Anderson wasn't much interested in wrestling, they talked him into it. He delivered a good old-fashioned Scottish "bonnie hank" and the Sumo wrestler found himself eating sawdust. Scottish fairness being what it is, the next encounter was conducted in genuine style. (The first had been a "Cumberland" match, similar to, but not identical with, Sumo wrestling.) As Webster describes it, "The Oriental launched himself at the Scot like a bullet from a gun, but Bill stood feet wide apart and didn't budge. In fact, he fought back and heaved to get his opponent off the ground." The match took a bewildering course after that, finally ending in some sort of tie when the Grand Champion Wakanohana stopped the bout because the Sumo opponent had started desperately pulling at Anderson's kilt to save the match (perhaps believing that quasi-nudity was a *sine qua non* for anyone gracing the Sumo ring?). Of course, it is not surprising that awesome Bill would fare well in Sumo. As indicated earlier, Cumberland-style wrestling, similar to Sumo, is a regular part of the Scottish Highland Games. Add to that Anderson's own natural power, plus technical virtuosity honed from years of competition, and you've got a match.

At 6 feet, 2 inches and 280 pounds of muscle, Anderson can be counted on to excel in any battle of brawn. Most recently he finished second in Britain's "Strongest Man" competition aired by the British Broadcasting Company, as well as placing high in numerous other international events.

Now well into his 40's, he—like the great Ed Corney of bodybuilding, namesake Paul Anderson in Olympic weightlifting, and Al Oerter of discus fame— continues to ridicule the myth about strongmen going to seed in middle age. Bill was in Paris recently with David Webster, where for six consecutive weekends the mighty one tossed progressively heavier cabers until he successfully tossed the heaviest caber ever seen (heavier even than then legendary *Braemar* caber, generally reckoned to be the grandaddy of all). According to authorities, Anderson is virtually unbeatable as long as awkward, heavy weights are used (though he recently out-threw a French weight-tossing champion while he and Webster were in Paris). Additionally, he is well known and well respected by the Royal Family, having given them many hours of entertainment at the Scottish Games events at home and abroad.

Arthur Rowe shared first-place honors with Bill Anderson in the 1963 Games.   (David P. Webster)

For all that, Anderson is a fine citizen and emissary for his homeland—living proof that a good natural athlete can compete with the best drug-trained warriors, since Anderson has never touched anabolics in his life. But he is human. Maybe Webster captured his essence best when he wrote me thusly of Anderson: "He is an upstanding, clean-living athlete, non-smoker, not a womanizer but human to the extent that he enjoys our national drink. Although he may be the type to crawl across a dozen naked blondes to get to a bottle of whiskey, he can hold his liquor and only drinks in moderation."

12

The Stones of Strength

If wrestling goes back to the dawn of civilization, the hoisting of heavy stones goes back to the dawn of Man himself. For the activity is a time-slice of the eternal battle: *homo sapiens* against the force of Nature, struggling for their very survival, with no finely honed implements, no repeating rifles, no modern weaponry of any sort to sustain primitive man—only force of limb.

As a sport, however, stone-lifting is a later phenomenon. Records of ancient Greece clearly show that the activity was only then beginning to evolve into the modern forms seen today. Huge boulders have been found inscribed with the names of the strongmen who supposedly lifted them. Other records indicate that the practice of pitting gristle against granite persisted throughout Europe well into the time of the philosopher William of Ockham in the late medieval period. In the Apothekerhof castle in Munich, there is an enormous rock weighing in excess of 400 pounds on which it is written that the Duke of Bavaria proved his manhood by hoisting it. And, though there is comparatively little by way of organized stone-lifting in Europe today, the medieval tradition exists in occasional informal competitions. The great Austrian bodybuilder, Arnold Schwarzenegger, won a number of such competitions when he succeeded in hoisting a famous rock in Munich: The rock weighed more than 500 pounds! Not to be outdone by the German tradition, the neighboring Swiss staged their own version of rock-lifting. Though this too has disintegrated somewhat in recent years, with contests being staged irregularly, the *Unspunnenfest* festival near Interlaken would regularly include an event consisting of tossing a 184-pound boulder (the *Unspunnen* stone). This unwieldy oval rock is nearly two feet long and over a foot in diameter. The object is to raise the stone to the chest and ultimately to the top of the head. The athlete then throws the rock as far as possible. Though unlikely to be mistaken for a meteorite, the rock can sail as far as 11 feet in the hands of a worthy strongman.

The Scots, of course, with their *clach cuid fir*, or "manhood" stones, are surely the dominant grapplers of granite in the civilized world. The power-conscious Scotsmen pride themselves on their nearly unliftable

The stones of strength have provided a challenge to lifters over the centuries. (David P. Webster)

Dinnie stones as well as the renowned Inver stone. It is said that only the very strongest can hoist the Dinnie stones a few inches off the ground. Even fewer succeed in raising the Inver stone to waist level. The mystique of the stones has, however, declined a bit in recent days, with the arrival of weight-trained American athletes on the bonnie shores. The great Bill Kazmaier, powerlifter *sans pareil*, actually succeeded in ramming the awesome 268-pound Inver stone over his head! No man in history had ever done that before. Some will quibble. They'll say that the Inver stone isn't the 562-pound barbell Alexeev lifted in Montreal; yet the awkwardness of that rock makes such comparisons futile. Whether Alexeev could duplicate Kazmaier's feat, even supposing the Soviet colossus could maneuver the Inver stone over his titanic tummy, is far from obvious. Noteworthy also is the fact that the great Scottish strongman, A.A. Cameron, in the early part of this century was one of the few who succeeded in lifting the stone a respectable distance. In the 1959 competition held at the Spartan Club of Aberdeen, most of the contestants, all prodigiously powerful, failed even to nudge the stone.

For a little international flavor, the exploits of the powerful French-Canadian Louis Cyr demand recounting. When Cyr was at the peak of his power, he made the legendary pilgrimage to Scotland to challenge the great Donald Dinnie himself. Locks flowing and muscles bulging, the great Cyr not only matched Dinnie pound for pound, but actually carried the stones farther than Dinnie did! Awestruck by Cyr's performance, Dinnie was moved to announce, "Man, you should hae been a Scot." To this the great strongman calmly retorted, "Monsieur, I am proud to be a French-Canadian." It must be remembered, of course, that Dinnie was over 60 years old when he meet Cyr in battle royal.

So great is the legend of the Dinnie stones that even the space age has entered the picture. British weightlifting champion Dave Prowse, better known as Darth Vader of *Star Wars*, has made the trek to the Dinnie zone. In October 1963, on a "lovely Autumn day," as Webster portrays it, dauntless Dave attempted the Dinnie stones before the TV camera and a large crowd. At first, possibly as a warm-up, he attempted each stone individually. No problem. But could he hoist both mountains of granite? According to Webster, "He felt sure he could. The wind made the tall pine trees wave, so Dave kept his track suit on and did not pause too long before he made his big attempt. Grasping the thin, rusty handgrips, he took a deep breath and slowly straightened his legs and back. Although there was quite a crowd of sightseers around by this time, you could hear a pin drop and there was a spontaneous gasp as the first boulder cleared the ground and a fraction of a second later the big one rose too."

There are other "manhood" stones in Scotland worthy of note. The *clach deuchainns*, or trial stones pose a worthy problem for any strongman in the world. Old "Black Donald," it is reported, used to lift the gigantic *Lochaweside* stone daily merely for exercise, to keep him in shape for twisting legs off of cows. And in Balquhidder near the Kirk there is the *Puterach* stone, which has resisted an infinite number of attempts to raise it barely to breast height.

Can any account of stone-lifting be complete without mentioning the great and noble Basques? A most fascinating and mysterious people, born into a tradition of strength and machismo, these hardy people of the Pyrenees of southern France and nothern Spain remain

Lifting champion Bill Kazmaier hoisting the Inver Stone. (David P. Webster)

one of the great enigmas of the 20th century. Their origins are shrouded in legend and uncertainty: Some claim that the language is so unlike anything known on earth that they must be descendants of the lost continent of Atlantis. Other, slightly more guarded accounts of the race, hold that theirs is a Mediterranean tongue, akin to Etruscan, and the original language of the Iberian people.

As with Japanese Sumo wrestling, the great tradition of Basque strength lore cannot be understood without some knowledge of their philosophies and cultures. One almost never sees women participating in the Basque stone-lifting events, a phenomenon readily explained by the male-dominated nature of Basque society. Also, stones play a more important role in Basque life than as mere objects of sport. In the quiet village of Ustaritz, the local parliament—the *Bilcar*—met regularly in days long gone. The assembly meetings were held on Sundays, usually around a giant stone symbolizing the power of the ancient Druidic origins of the Basque people.

More evidence of the great importance of enormous boulders in the Basque worldview is found in their fairy tales—always an authoritative source of cultural knowledge. In the tale of "The Stupid Tartaro," dear to every Basque child, there's an exemplary instance of Basque resourcefulness and fascination with sheer physical prowess. In a quasi-biblical confrontation, the boy of the story is destined to outwit the Tartaro, a giant of a man satanically bent on devouring him. The great battle begins with a test of strength. "And I eat boys like you!" the Tartaro roars. "We shall soon find out who is the strongest. Tomorrow we shall have a test of our strength and we shall see who can throw a stone to the greatest height. If you should win, I will not eat you just yet." With plucky resourcefulness, the boy tosses not a rock, but a bird, thereby fooling the giant into thinking he was indeed the more powerful, since the "stone" obviously flew on its merry way. Thus the boy saves his neck and demonstrates, incidentally, that while Basque lore reveres strength, it reveres character and intellect more.

As practiced today, the rules forbid the throwing of birds. The point of competition is to determine who can hoist the greatest tonnage over time. It is common for the winner of such contests to hoist nearly ten tons in the course of one day. Obviously, the endurance factor is important, demanding perhaps more than any of the other strength sports currently practiced. Combatants will often make over *one hundred* attempts at a lift, by contrast with their Olympic lifting brethren who are allowed three tries per lift. Little attention is paid to weight divisions in competition, often leaving the smaller man behind. According to some Basque enthusiasts I talked to, the practitioners regard the division into bodyweight classes as an uninteresting artificiality. ("Could our ancestors plead with Mother Nature to send a smaller Bull Mastodon, that we might defend ourselves with rocks more within our capabilities?")

As with so many of the ancient sports, the origins are shrouded in legend, mystery, and hearsay. Generally, experts agree that the earliest beginnings go back at least to the Middle Ages. According to one account, some Basque shepherds are said to have devised a lifting contest to alleviate their boredom. The practice quickly spread, finally involving whole towns warring against one another for strength supremacy.

The stones used in such battles are enormously heavy and unwieldy, much like the implements of the Scots. One such is the *easarone,* a cylindrical rock with two holes at either end for gripping. Such boulders often exceed 400 pounds and are as awkward as sacks of flour. The idea is to hoist the rock from the ground to neck height as many times in one minute as possible. One of the greatest in this event was, for many years, a sturdy Basque laborer named Ostolaza, who made a staggering 21 lifts in one minute! Having thus warmed up, he proceeded to make 263 lifts in a half hour. And with a "lighter" 275-pound easarone, the great Irazusta is reputed to have elevated it 15 times in two minutes. The rocks go heavier even than this; the giant Manterola muscled up a 330-pound easarone 178 times in less than a half hour as well, while Urtain, the fabulous European heavyweight boxing champion, did 186 lifts with the 220-pound easarone in less than 30 minutes. For the gigantic 440-pound rectangular rock revered by the best of the Basques, the mighty Aguirre is alleged to have conquered it seven times in five minutes.

The other type of rock used in the competitions is actually a granite cylinder, polished until smooth; these also vary extremely in weight. There seems to be no standard size used in Basque lifting events. There is one in Spain weighing more than 500 pounds, though it is little used. Whichever type of rock is adopted, however, the goal never varies: The granite must be raised as many times within a specific time period as possible. Usually a "period" consists of two half-hour rounds, with a short break between.

Because of the potentially damaging effects of the boulders, lifters generally wear garments—the *gurreko* and the *chaleco.* The latter is a vest protecting the upper body, while the former is functionally equivalent to the thick belts worn by the Olympic weightlifters and powerlifters, intended to provide support for the midsection. It provides, additionally, considerable leverage during the lifting.

From a technical point of view, the Basque lifts probably don't compare in the virtuosity and speed required by the Olympic-style lifts. However, the lifting of the cylinder requires much practice. Like the clean and jerk, this event involves two phases. First the lifter grips the cylinder at either end and immediately rams his stomach (which can be ample) into the top of the rock to form a turning point. He then turns the weight so that one end continues to rest on the gurreko. The contestant, now in an upright or slightly bent-back position, speedily releases his hold on the upper part, cradling the bottom with both hands. In the second phase of the lift he simply thrusts it onto his shoulder, parallel to the floor. At the referee's clap, he releases it slowly, for as in Olympic lifting, points can be lost for allowing it to drop to the floor.

There is an interesting variation on European Basque lifting popular among Basque residents of Idaho. An overpowering strongman named Juan Uberuaga is rumoured to have carried two 103-pound stones a distance of 230 yards; while in yet another variation of the great Basque mania, a rock-throwing contest attracts crowds from miles away. Martin Idoeta won a few of these by heaving 105-pound stones as far as 11 feet, while, in still another variation, Jose Maruri hoisted a 235-pound pipe to his shoulders 13 times in less than four minutes.

Whether tossing pipes in the United States or conquering the granite in the lovely villages of the Pyrenees, the Basques maintain the great natural and unaffected tradition of their culture, modeling their training after the actual competitions. No fancy "assistance" exercises unconnected with the final realities of the lifting of the stones, and no silly dietary fads to spoil the enjoyment of a robust Basque *paella*. The workouts are brief and intense, involving frequent rests as the actual competitions will demand. As seems inevitable, however, tales pop up here and there of some enthusiasts trying the fashionable American diets of peanut butter, high-protein milk shakes, and pork-liver casseroles.

Will this dietary heresy be tolerated? Maybe, if the American weightlifters can prove all that stuff makes them stronger than their Basque cousins. But none of the Basque grocers are holding their breaths.

Benito Goitiandia

If there is one name that shines in Basque lifting like the great Spanish moon itself, it is that of Benito Goitiandia. Huge, agile, guardedly friendly, and supremely confident of his prowess, the big man has been compared to the great Alexeev himself in strength and appearance. Weighing nearly 250 pounds, he has dominated the sport in the Pyrenees for years. He is, allegedly, the only man in Spain to have lifted the great granite ball of more than 550 pounds. In fact, he is on record as having powered up the recalcitrant orb seven times in less than three minutes. Still another of the great Basque champion's feats consisted in hoisting the 225-pound cylinder to shoulder height, rolling it across his back and onto the other shoulder. At that point, when all expected the usual release of the rock to the floor, he rolled it across his massive chest, back to the starting shoulder, finally completing the dizzying cycle three times!

And, as the comparison to Alexeev shows, the temptation to compare the Basque supermen with the Soviet lifters is immense. How would Goitiandia fare against Alexeev? Fortunately or unfortunately, depending on your perspective, we have not and most likely will not have any such trial. What is known is that a weightlifter of some considerable reputation (who shall, mercifully, go unnamed) did take part in a rock-lifting contest against Basque strongmen far inferior to the great Benito. While the other contestants averaged 60 to 90 successful lifts, the barbell-trained enthusiast salvaged precious self-esteem in making nine good lifts.

For all that, devotion to the Basque mania could well improve such performances. But full equality with the Basques? Never mind that the question borders on sacrilege; the point is that the Basques and their rock-raising colleagues in Germany and Scotland probably don't give a damn about such matters. What remains is the hard fact that stone-lifting and the likes of the fabled Goitiandia in the rolling hills of the sun-baked Pyrenees will remain to enthrall generations of strength lovers, schoolchildren, and sports fans everywhere.

13

The Muscular Men and Women of History

In previous chapters we explored the glorious, glamorous universe of muscle in all its variety. However, that universe has not always been so precisely defined, nor taken so seriously as it is today. The roots of modern weightlifting, at least in America, Russia, and a good many other parts of the world, go back to vaudeville: no rules, exacting regulations, or governing bodies. Tricks and dangerous stunts were the order of the day.

As in many parts of the muscle world, it is often extremely difficult to separate fact from fiction. Did the mighty Paul Anderson really push a loaded freight car up an incline, or hold a twin-enging DC3 back from take-off? Perhaps the fabled Katie Sandwina really did hurl a 300-pound man 20 feet through the air. We may never know for certain. Still, through oral tradition and the magic of photography, we have some reason to believe many of the mystifying feats of power really did take place. The historic roster of strongmen is long, and cannot be fully covered in a chapter. But we recount the greatest.

Louis Cyr

There will be little quarrel with the choice of Louis Cyr for these pages. Born in St. Cyprien de Naperville in Canada in 1863, he is considered by many to be *the* strongest man who ever lived. If the supporters of Cyr are to be believed, he would have mocked Anderson, put Alexeev to eternal shame, and humiliated the feats of Reinhoudt. Closer to reality, Cyr was virtually born into immense strength. Very early in his career he was employed in the Merrimack Mill in Lowell, Massachusetts, a small industrial town that even today looks like a picturesque hamlet in up-state New York. The character and climate were right for Cyr. He was then all of 14 years old—too young to begin his career as a strongman, but old enough to acquire a job that would begin to elicit his awesome abilities. (It was widely held in those days that niceties such as a formal education were a luxury for working-class French-Canadians.)

According to George F. Jowett in his monumental work on Cyr, *The Strongest Man that Ever Lived*, an incident took place the day Cyr arrived in Massachusetts that supposedly began what later turned out to be one of the remarkable careers in muscledom. As it is told, the foreman in Lowell was something less than overwhelmed by the young Cyr. At exactly 5 feet, 1 inch of height and a mere 170 pounds, Cyr was far from an Anderson. According to Jowett, the foreman's only comment was, "He's too fat; could be slow and lazy." Cyr was prepared: "Try me, I'm very strong," he retorted. Without waiting for a response, young Cyr promptly hoisted a 150-pound bale of cotton and linseed onto his shoulder. Cyr got the job.

He worked at the mill for the next two years, gradually developing the monstrous physique and power that would eventually carry him to the summit of sinewy strength. Though lacking the much-touted benefits of modern Nautilus machinery and chrome-sleeved barbells, Cyr made do with hours of pushing hand trucks laden with weighty bales of cotton, flax, string, and wood. Though he grew considerably during his tenure at the mill, the generous eating habits of the young Hercules kept his massive muscularity hidden under rolls of fat. As a consequence he gave virtually the same impression when he tried to hire himself out as a farmhand. Lowell farmer Dan Parker was less than impressed, to say the least, regarding Cyr as too fat and slow for the job. Still, at the kind urging of Mrs. Parker, Cyr was given a job once again, and in keeping with history's eternal cyclical nature, it was another feat of prodigious power that won the boss's favor. Parker was planning to build a new wall and had put Cyr to work loading the truck, with orders to have it hitched up to the horse and pulled down the road. Predictably arrogant, Cyr lugged the enormously heavy cart down the road himself! (One is reminded here of Scottish strongman Donald Dinnie embarrassing his horse by dragging a loaded wagon out of the mud alone.)

After that, young Louis could do no wrong. He lived with the Parkers for two more years, a virtual son to Dan Parker. By this time he had grown to a full 5 feet, 8 inches and a weight of 250 pounds. The itch to astound the world with his apocalyptic strength was beginning to stir. Opportunity was not long in knocking. A common stunt in those days was the lifting of a full-grown horse—a feat only the mightiest dared attempt. The rules were simple enough: The horse had to be hoisted completely off the ground without the assistance of mechanical devices of any sort. In one such competition, the other competitors failed to budge a particularly well-fed stallion, whereupon Cyr talked the skeptical manager of the contest into allowing him a go at it. What happened was truly biblical. As the crowds jeered at Samson as he placed himself between the pillars, so the crowd taunted Cyr on that sunny day in Middlesex County. They laughed at him as he approached the horse. They laughed at his ungainly position underneath the immense beast. They laughed still more as he took his first deep breath of life-giving oxygen.

Louis Cyr, perhaps the strongest man who ever lived. (Public Archives Canada C 86343)

Then the laughing stopped. Slowly, inexorably, the horse and Dame Gravity fell before the onslaught of the Canadian superman. The thunderous ovation the crowd gave Cyr can, some will claim, still be heard echoing across the rolling Canadian countryside. So overwhelming was the approval that Cyr knew his destiny was to become a professional strongman. The choice was logical. In the economically chaotic days of 19th-century North America, a professional lifter was a respected and highly-paid performer. He would remain so until the opening days of the First World War.

Jowett recalls an incident where Cyr, during a brief flirtation with a career as a policeman, was attacked by a gang of 25 muggers. Most survived, but the incident convinced Cyr that police work was quite boring. Although there is no official record of Cyr's ever being defeated, a story persists where a woodsman from Michigan beat him in an off-hand lifting contest. Most likely the story will never be validated, though French-Canadians still love to debate it.

What Cyr is known to have done officially is documented in Jowett's book. He's on record as having demolished such distinguished strongmen as Romulus, Biekowski, the terrifying Cyclops, Big John "Ajax" Whitman, Horace Barre, Donald Dinnie, Dave Michaud, August Johnson, Richard Pennell, and the legendary Sebastian Miller. En route to such victories, Cyr back-lifted 4562 pounds, one-arm pressed 243 pounds, two-arm pressed 347 pounds, one-hand deadlifted 987 pounds, one-finger lifted 554 pounds, one-arm "swung" 174 pounds, and executed countless horse lifts and bag lifts. His power was so unbelievable that Richard Fox, editor of the *Police Gazette*, issued a challenge, offering $2500 and a $1,000 diamond-studded belt, plus the title of "Champion of the World," to anyone in Europe who could best Cyr. No one ever undermined Fox's faith in Cyr, and the treasures remained in Cyr's possession until his death 20 years later. Eventually the belt was bequeathed to the worthy Warren Lincoln Travis for his many feats of strength. It rests now in the Weightlifting Hall of Fame in York, Pennsylvania.

In his prime, Cyr weighed 365 pounds and stood 5 feet, 10 inches tall. Some claim he had biceps over 22 inches around, 19-inch forearms, and a 59-inch chest (unexpanded). Such girth was extremely difficult to fathom in those pre-steroid days, though there is no question of Cyr's being equal to his reputation. When he died on November 10, 1942, flags were flown at half-mast and he was mourned throughout Canada. Possibly the equal of today's best, Cyr was one of the most loved, influential, and respected of all strongmen in history.

Hermann Goerner

It is a long trek from the beautiful Canadian hills to the magical forests of Germany, but no account of historic strongmen can presume completeness without some mention of the great Hermann Goerner. Though pitifully weak as a child, he was already far, far stronger than two average men by his teens. Born near Leipzig in 1891, Goerner stood over 6 feet and tipped the scales in excess of 260 pounds. Goerner was introduced by the entrepreneur W.A. Pullum to the British public in the late '20s. In 1927, J. Paul Getty paid Goerner almost $1000 for making a right-hand deadlift with 602 pounds—ample testimony to the enormous strength and size of Goerner's hands, as well as his back and legs. Despite his world travels, Goerner remained a hero to the people of Leipzig. According to his biographer, E. Mueller, "It need hardly be stressed that Goerner was Leipzig's greatest attraction in 'Iron Game' circles . . . It is no wonder that Leipzig and other weightlifting clubs engaged Goerner for their Exhibition Shows, knowing full well that 3000 to 4000 spectators would show up to see Goerner in the flesh . . ."

Part of Goerner's striking uniqueness as a professional strongman is, according to biographer Mueller, the fact that many of his greatest feats of strength were performed after age 40. Besides his many records in the more or less conventional right-hand snatches, where he lifted nearly 230 pounds, and a right-hand clean of over 297 pounds, Goerner performed an astonishing number of offbeat lifts that characterized the vaudevillian nature of early weightlifting. He is renowned for having jerked a human barbell weighing nearly 400 pounds—a feat made even more incredible in that the live humans at either end presented a formidable balance problem. (Alexeev might well have trouble jerking that weight overhead today.) Possibly more remarkable is Goerner's feat of carrying a grand piano weighing over 1400 pounds for a distance of over 50 feet. He is known to have carried a half-ton of bricks up a flight of stairs, raised a car by the front axle, and out-muscled six pro wrestlers in arm-wrestling in less than a minute. At the supposedly advanced age of 58, he is known to have smashed a Colin hand dynamometer with the strength of his grip—a feat requiring gripping power in excess of 286 pounds. And according to Terry Todd, his right-hand deadlift of 727½ pounds is the greatest single feat of strength ever recorded by a modern strongman. Mueller summed up the great strongman's career in one simple, yet profoundly moving, utterance: "The world will never see the likes of Goerner again."

Katie Sandwina

The audience is silent. The lifter approaches the immense load of iron as a wild beast to be tamed. The cannon, 1200 pounds of sheer dead tonnage, has defied the finest efforts of the world's strongest. Within seconds the lifter bends to grasp the great weight, rises, and walks several feet with the half-ton weapon slung across the shoulders. As effortlessly as it was hoisted, *she* tosses the cannon insultingly aside, as if it were a mere nuisance to be promptly dismissed.

No misprint. The gentle lady, Katie Sandwina, was surely the most impressive female figure in the annals of strength. A shapely woman, despite her 6-foot, 200-pound frame, Katie sneered at the laws of biology by performing feats of power that eluded even the best strongmen of her time. As tradition has it, Katie is thought to have inherited her enormous natural force from her father, Philip Brumbach, himself descended from a venerable line of supermen. Born in a wagon in a gypsy caravan in Germany, circa 1884, she was destined to crisscross the globe, performing before monarchs and emperors. She reached her all-time peak around the turn of the century, a glorious and opulent time for weightlifting heroes.

The origin of Katie's name has caused considerable controversy in strength circles. And, as these things go, it is once again difficult to parse out truth from exaggeration. Some authorities claim that Katie actually defeated the great strongman Eugene Sandow, whence she forever adopted and adapted his name, billing herself as "Sandwina" in her theatrical performances.

One of the most discussed anecdotes in Katie's career involves a local strongman who was not, apparently, in total sympathy with the ideals of women's liberation. As the tale is told, Katie declined (as she always did) to participate in a contest of strength with an aspiring local superman. Eventually, however, the taunting and ribbing overcame her and natural ego took over. She went over to the troublemaker, a chap approximating the girth of a small automobile, whisked him overhead for a full minute, and lazily jettisoned the fellow into the less than eager arms of his friends, all of whom collapsed under the weight.

Katie's sense of showmanship being what it was in incidents like the above, it's not surprising a good deal of her career involved circus work. It is said that Katie performed her strength act at Barnum & Bailey Circus just a few hours before the birth of her first child (who, incidentally, went on to uphold the great name of Sandwina by becoming boxing champion of Europe). Stunts she performed specifically for the circus included holding two teams of horses apart with arm and shoulder power alone, supporting a bridge across which 40 men and four horses rode, and shouldering a 1200-pound cannon. In my judgment, only Ann Turbyne's deadlift of 450 pounds can compare with such feats in the history of woman's weightlifting. Indeed, Katie demonstrated her staying power by breaking heavy chains and bending iron bars at well past 60 years of age.

There have, of course, been other strong women. The stupendous "Minerva" comes to mind—a Hoboken housewife who toured the United States performing an impressive routine. It was not sufficient, however, to displace Sandwina in the record books. Somewhat later, Marta Fara toured Europe billing herself as the "Strongest Woman in the World." (Many strongmen have lost considerable credibility in both past and present years by billing themselves as "The Strongest Man in the World." Though the device worked a little better in the chaotic vaudeville days, it's foolish to claim the designation today unless your name is Vasili Alexeev.) Fara was quite good, supporting over 4000 pounds in the "Tomb of Hercules" position and harness-lifting an elephant.

As with the male superheroes of strength, the temptation to compare the ancients with today's champions is irresistible. How might Ann Turbyne or Jan Todd have fared against Katie Sandwina? If the history of the matter is anything approaching accurate, it would seem that today's iron amazons might have a difficult time. That even Ann Turbyne could shoulder a 1200-pound cannon is hard to believe, to say the least. Of course, there is no official verification of the weight of the weapon, though they're anything but light. (I, for one, would be most interested in seeing today's strongest women attempt to duplicate Sandwina's feats—a remote, but theoretical possibility.)

Smaller—but Stronger!

To this point we've mentioned only the goliaths of muscle, be they male or female. Axioms of fairness demand that the smaller, but nonetheless remarkable, strongmen be allowed their place in the sun. To be sure, there does seem to be a greater immediate drama in watching the biggest humans hauling up the biggest weights, for whatever reasons that reside in the human psyche. Yet, as is well known today, the smaller men often lift proportionately greater weights than the behemoths. Where, for example, can one find a better pound-for-pound Olympic lifter than the amazing David Rigert of the Soviet Union? Though nudging the scales at only slightly over 200 pounds, Rigert snatches weights comparable to what Alexeev snatches, though the gargantuan teammate of Rigert's weighs in at better than 350 pounds! And, though many have mouths far bigger than Rigert's, there is not a single superheavyweight or heavyweight lifter in the United States who would not be utterly demolished in a contest with the Russian. (Of course, the likes of Lee James, bronze-medal winner in Montreal, and Mark Cameron, ace 242-pound lifter, approach Rigert in caliber.) And Rigert is only one example of the truly awesome power exhibited by smaller men. The "Mighty Atom," Joe Greenstein, who recently verified one of the old sayings about strongmen by dying "young" at age 96, is the top choice for many. When in his prime (around age 75 or so), he repeatedly pulled cargo trucks down Broadway with his teeth, broke chains, and drove spikes into boards (though many of these stunts were continued well into his 90s!).

Not to be outshone, Dr. E. Petulick, a dentist from Rochester, New York, lifted 500 pounds with his teeth (proving something, one would suppose, about the advantages of daily brushing). As this bizarre style of lifting became increasingly popular, the great Warren Lincoln Travis bit into the task of finding a place for himself in muscle history by holding a small merry-go-round suspended from a chain fastened to a bit locked firmly between his jaws.

In Europe Hans Streyer, another famed oldtimer and marvel of muscle, lifted a quarter-ton with one finger while holding a 55-pound weight at arms' length, again with only one finger. And many other renowned old-time strongmen have excelled in finger strength without serious specialization. Arthur Saxon, though usually remembered for his superb 370-pound bench press, is alleged to have raised 297 pounds overhead using only his little finger. The weight was hooked through a leather loop, through which Saxon slipped his finger. Also, Sandwina's father, Philip Brumbach of Munich, is credited with hoisting 642 pounds with his middle finger alone, as well as effortlessly raising 441 pounds with the little finger in 1880.

Among the more imaginative strongman stunts was the card-tearing of Al Treloar. Witnesses claim that he would regularly tear in half three decks of playing cards at every performance of his vaudeville career. Additionally, he bemused the audience by juggling 16 chorus girls in the back lift, followed immediately by a jaunt up a ladder with a full-grown horse on his shoulders.

For doing the most to outlive his name, the first prize must go to Mac Sick of Bavaria. For the sickly, runty child from Europe became, eventually, the first man in history to shove double bodyweight overhead (332 pounds). By virtue of his staggering strength exploits, hauntingly anticipating the future accomplishments of such monumental heroes as Berger, Miyake, and Vinci in the lighter bodyweight classes, he was selected to journey to Great Britain under the auspices of the South African promoter Tromp Van Diggelen. Dissatisfied, for obvious reasons, with his protégé's given name, Von Diggelen promptly changed it to "Maxick." There the legend began. Despite his 145-pound frame, Maxick pressed 230 pounds, snatched 220 pounds, and clean and jerked 211 pounds with one hand.

The list of strongmen in the heady days of vaudeville is virtually infinite. Otto Arco, Nordquest, Lambert, Rockstool, Hoffman, Jowett, Klein, Matysek—all are worthy of mention. Yet time moves on, strongmen die, and new heroes mount the throne of Olympus. From the '20s through the '50s there was nary a star quite equal to the likes of Cyr or Goerner. But in the mid-'50s a giant appeared whose name would be forever synonymous with immortal strength. Paul Anderson had arrived. Though his great name has been mentioned several times throughout this book, his exploits are impossible to overwork. So awesome, so electrifying was his prowess, that the Russians themselves dubbed him a "wonder of nature." The great "Dixie Derrick" completely demolished all the world's records in the heavyweight class before retiring from competition to run a boy's home in Georgia that he continues to operate to this day. He is said to have squatted with well over 1100 pounds on his back, pushed freight cars, twisted gigantic spikes—all child's play for Anderson. Though his untimely death was predicted everywhere in the '50s because of his immense bulk, Anderson continues to sneer at the skeptics—most recently by performing six full squats with over 675 pounds on his shoulders at nearly fifty years of age. And despite his age he shows no signs whatsoever of declining prowess. A dedicated evangelical Christian, Anderson regularly delivers the gospel of Christ to youth groups across the country. Nor does he find any contradiction in his lifelong preoccupation with both body *and* spirit. He considers his Christian "rebirth" part and parcel of his unbelievable feats of strength. Indeed, he regularly incorporates awesome displays of might into his missionary work, juxtaposing a fiery condemnation of atheistic and materialistic values with the torpedoing of a gigantic spike through a solid piece of wood.

Though it is unlikely the universe will soon bear witness to another phenomenon such as Anderson in the very near future, there are worthy aspirants to the accolades of Atlas even today.

How could one fail to mention the mighty Franco Columbu? At 187 pounds, he may well be the strongest man in the world for his size. Lifting automobiles, blowing up hot-water bottles, and bending thick steel spikes are all well within the Sardinian's abilities. In *Coming on Strong*, he describes his most popular stunt, the stupendous hot-water-bottle feat. As he explains, "But my most popular stunt by far is blowing up a hot-water bottle like a balloon till it bursts . . . the bag will be nearly the size of my chest before it goes, maybe 20-odd breaths, and the audience will be far more tired than I by then, straining all that time in fear of nothing more than a loud noise, one of the fears we are all born with." (Franco warns, incidentally, that under no circumstances should anyone try this, as the air in the bag can utterly destroy your lungs if you let go, the return pressure can reach 500 psi.) In addition to

Superhuman Paul Anderson uses his strength to spread faith in God. (Paul Anderson Youth Home)

Lifting a car is all in a day's work for Paul Anderson. (Paul Anderson Youth Home)

these spectacular stunts, Franco has officially bench-pressed 475 pounds, squatted with 655 pounds, and deadlifted 745 pounds, making him one of the great powerlifters in the world.

Speaking of incredible wind power, one need only look at the following claim by strongman Mike Dayton: "I could pull down the pillars of Mount Olympus, if they were still standing. There are no limits or restrictions on my strength. I have done and continue to do feats of strength which doctors, bodybuilders and powerlifters say are impossible. For me there is nothing which is impossible." In a similar incredible vein, Dayton claims to be able to withstand a six-foot drop from the gallows and a 212-foot jump from the Golden Gate Bridge. Whew! Given the supermen we've covered in the book, it seems extremely unlikely that Dayton is as strong as he claims. Indeed, he was beaten by several strongmen in CBS's "World's Strongest Man" competition. What Dayton is expressing, I believe, is the phenomenal power of mind control—the unique ability of the mind to harness normally unused reserves of power.

The philosophy Dayton espouses is called *Ch'I*, one of the many variations of Eastern metaphysical thought. Of course, western philosophers have long since shown the inadequacies of Eastern thought—at least as a system claiming to express literal truths about the world. Actually, Ch'I is simply another variation of the old idea of mind over matter.

Reduced to its essentials, Ch'I stresses the fact that the mind can accomplish extraordinary things if one can muster sufficient concentration and mental discipline. This, of course, is correct. Though the ideas are hardly complex or sophisticated enough to warrant so austere a designation as "philosophy," the basic idea does work. Indeed, stories abound where ordinary people have accomplished striking physical feats in extraordinary situations. The classic story, never actually verified, is the one where an old woman lifted an automobile to save a child. Another such is the ability of some Eastern mystics to withstand a walk over hot coals with bare feet. The latter is certainly true, and the former story may be true. Dayton is quite right that the mind plays an extremely important part in feats of strength, and some sort of subconscious use of the idea of Ch'I may well explain the ability of the old woman to muster auto-lifting strength. Dayton himself has broken police handcuffs—a feat that one would not expect any human being to be capable of, no matter how devoted a follower of the Iron Game. Also, the Soviets are no strangers to the art of mind control—call it Ch'I, hypnosis, Zen meditation, Yoga. Fabled David Rigert outstrips all but the finest superheavies, though he weighs less than 220 pounds. Alexeev hoists more than 500 pounds and lifters are breaking the 500-pound "barrier" all over the place. The breaking of psychological barriers plays no small role in the sports of strength.

The question: How far can it all go? What are the uppermost absolute limits of human strength? In my judgment there is no question that someone, somewhere, will hoist 600 pounds over his head. And 700? Who knows? Chemistry plays an important role in the contemporary sport scene, regardless of one's ethical feelings on the matter. And given the history of prior predictions regarding the limits of human potential, it would be patently absurd to speculate very far. Obviously no human being will ever succeed in lifting the Rock of Gibraltar. Yet it is my feeling that the 700-pound clean and jerk is within the limits of conceivability; it may well be the upper threshold of human potential and will surely take some colossal chemical android/human of the future to accomplish this. But such musing becomes empty after a while. This chapter was about the strongmen of history, and before we have to allow for the advent of steroid-injected monstrosities of the future, let us dwell for a final moment upon the incredible accomplishments of the likes of Sandow, Sandwina, Cyr, and Anderson. We shall never see them or their like again.

Appendix

BODYBUILDING — MEN AND WOMEN

Governing Bodies

a) International Federation of Bodybuilders (IFBB)
   2100 Erwin St., Woodland Hills, CA 91364
b) Amateur Athletic Union of the U.S. (AAU)
   3400 W. 86 St., Indianapolis, IN 46268
c) World Bodybuilding Guild (WBBG)
   1665 Utica Ave., Brooklyn, NY 11234
d) Superior Physique Association (SPA)
   P.O. Box 937, Riverview, FL 33567
e) American Federation of Amateur Bodybuilders (AFAB)
   P.O. Box 662, Holyoke, MA 01040

Equipment Sources

a) York Barbell Co.
   York, PA 17405
b) IFBB (see above)
c) WBBG (see above)
d) Iron Man's Body Culture Equipment Co.
   Box 10, Alliance, NB 69301

Bibliography

a) Columbu, Franco, and George Fels. *Coming on Strong*. Chicago: Contemporary Books, 1978

b) Fallon, Michael, and Jim Saunders. *Muscle Building for Beginners*. New York: Arc Books, 1967

c) Schwarzenegger, Arnold. *Bodyshaping for Women*. New York: Simon and Schuster, 1979

d) Kennedy, Robert. *Bodybuilding for Women*. Buchanan, NY: Emerson Books, 1979

e) Columbu, Franco, and Anita Columbu. *Starbodies: The Women's Weight Training Book*. New York: E.P. Dutton, 1978

f) Barrilleaux, Doris, and Jim Murray. *Inside Weight Training for Women*. Chicago: Contemporary Books, 1978

g) Gaines, Charles, and George Butler. *Pumping Iron*. New York: Simon and Schuster, 1974

h) Coe, Boyer, and Bob Summer. *Getting Strong, Looking Strong: A Guide to Successful Bodybuilding*. New York: Atheneum, 1979

OLYMPIC WEIGHTLIFTING

Governing Bodies

a) Amateur Athletic Union of the U.S. (AAU) Weightlifting Division
3400 W. 86 St., Indianapolis, IN 46268

Equipment Sources

a) York Barbell Co.
York, PA 17405

b) Iron Man's Body Culture Equipment Co.
Box 10, Alliance NB 69301

Bibliography

a) Murray, Jim. *Inside Weightlifting*. Chicago: Contemporary Books, 1978
b) Columbu, Franco and Richard Tyler. *Winning Weightlifting and Powerlifting*. Chicago: Contemporary Books, 1979
c) Fodor, R.V. *Competitive Weightlifting*. New York: Sterling Publishing Co., 1979

POWERLIFTING – MEN AND WOMEN

Governing Bodies

a) Amateur Athletic Union of the U.S. (AAU) Powerlifting Division
3400 W. 86 St., Indianapolis, IN 46268 *or*
Joe Zarella, Chairman AAU Powerlifting Division
P.O. Box 43, Hudson, NH 03051

Equipment Sources

See Olympic Weightlifting

Bibliography

a) Columbu, Franco, and Richard Tyler. *Winning Weightlifting and Powerlifting*. Chicago: Contemporary Books, 1978
b) Todd, Terry. *Inside Powerlifting*. Chicago: Contemporary Books, 1978

c) Hatfield, Fred C. *Powerlifting: A Scientific Approach.* Chicago: Contemporary Books, 1981

SUMO WRESTLING

Governing Bodies

a) Japan Sumo Association
   Tokyo, Japan

Equipment Sources

a) Japan Sumo Association
   Tokyo, Japan

Bibliography

a) Kuhaulua, Jesse, and John Wheeler. *Takamiyama: The World of Sumo.* New York: Kodansha International USA, 1973

Also contact: Sekai Bunka Publishing, Inc.
501 Fifth Ave., New York, NY 10017

Japan Publications Trading Co.
1255 Howard St., San Francisco, CA 94130

Kinokuniya Book Store of America, Ltd.
1581 Webster St., San Francisco, CA 94115

Charles E. Tuttle Co.
28 South Main St., Rutland, VT 05701

Embassy of Japan
2520 Massachusettes Ave., Washington, DC 20008

WRESTLING

Governing Bodies

a) United States Wrestling Federation
   4000 W. 19th St., Stillwater, OK 74074
b) Federation Internationale des Luttes Amateurs
   12 Valmont, Lausanne, Switzerland
c) United States Wrestling Foundation
   Wrestling Division, AAU, 3400 W. 86th St., Indianapolis, IN 46268
d) National Wrestling Coaches Association
   c/o Athletic Dept., University of Utah, Salt Lake City, UT 84112
e) Eastern Intercollegiate Wrestling Association
   P.O. Box 3, Centerville, MA 02622

Equipment Sources

a) AAE Equipment Co.
   4 Portland St., W. Conshocken, PA 19428
b) Goldner Associates, Inc.
   Dept. T Box 2703 63 Arcade, Nashville, TN 37219

Bibliography

a) Valentine, Tom. *Inside Wrestling.* Chicago: Contemporary Books, 1972
b) Keen, Clifford P., et al. *Championship Wrestling.* New York: Arco Publishing Co., 1973
c) Columbu, Franco, and Dick Tyler. *Weight Training and Body Building: A Complete Guide for Young Athletes.* New York: Wanderer Books, 1979
d) Clayton, Thompson. *A Handbook of Wrestling.* A S Barnes, San Diego: 1973
e) Columbu, Franco, et al. *Weight Training for Young Athletes.* Chicago: Contemporary Books, 1979

HAMMER, DISCUS, SHOTPUT, JAVELIN

Governing Bodies

a) United States Track and Field Federation
30 N. Norton Ave., Tucson, AZ 85719
b) Amateur Athletic Union of the U.S. (AAU)
3400 W. 86 St., Indianapolis, IN 46268
c) Track and Field Association of the USA
Kansas City, MO 64153

Equipment Sources

a) Saucony Shoe Mfg. Co., Inc.
12 Peach St., Kutztown, PA 19530
b) Gill Track and Field Equipment
201 Courtesy Rd., Urbana, IL 61801
c) Pacer Track and Field Equipment
715 Industrial Pkwy., Carson City, NV 89701

Bibliography

a) Bush, Jim, and Tom Valentine. *Inside Track.* Chicago: Contemorary Books, 1974
b) McWhirter, Norris D., and Ross A. McWhirter, eds. *The Guinness Book of Olympic Records.* New York: Bantam Books, 1980
c) Menke, Frank G., *Encyclopedia of Sports.* New York: Doubleday, 1977
d) Doherty, J.K. *Track and Field Omnibook.* Tafnews Press, 1980
e) Plimpton, George. *Sports!* New York: Harry N. Abrams Inc., 1978
f) Norback, Craig, and Peter Norback. *The New American Guide to Athletics, Sports and Recreation.* New York: New American Library, 1979
g) Wilt, Fred, ed. *The Throws: Contemporary Theory, Technique and Training.* Tafnews Press, 1980
h) Ecker, Tom. *Track and Field Dynamics.* Tafnews Press, 1979

WRIST AND ARM WRESTLING

Governing Bodies

a) International Arm Sports Association (IASA)
   P.O. Box 42070, Houston, TX 77042
b) World Professional Wristwrestling Association, Inc.
   P.O. Box 247, Duluth, GA 20137
c) World Wristwrestling Association
   830 Petaluma Blvd., N. Petaluma, CA 94952

Equipment Sources

Contact above

Bibliography

a) *Sports and Games Almanac.* Track and Field Association of USA, Kansas City, MO 64153
b) Kirkley, George W. *Weightlifting and Weight Training.* New York: Arc Books, 1966

SCOTTISH HIGHLAND GAMES

Governing Bodies

a) Royal Braemar Highland Society
   c/o David Webster, 43 West Rd., Irvine, Ayrshire, Scotland

Equipment Sources

For general strength-building equipment, *see* Bodybuilding—Men and Women. For equipment specifically related to the games, contact Royal Braemar Society

Bibliography

a) Webster, David. *The Scottish Highland Games.* Brooklyn: Beekman Publishers Inc.

STONELIFTING

Governing Bodies

a) Elko Euzkaldunak Club (for Basque Stonelifting)
   Elko, NV 89801
b) Embassy of West Germany (German stonelifting)
   4645 Reservoir Rd., N.W. Washington, DC 20007
c) Embassy of Spain (Basque lifting)
   4200 Wisconsin Ave., N.W. Washington, DC 20016
d) Deutscher Sportbund, Otto-"Fleck-schreise 12 (German stonelifting)
   6000 Frankfurt, am Main 71, W. Germany
e) Institute fur Sportwissenschaften der Universitat Wien (German Stonelifting)
   7800 Karlsruhe, den Postfach 6380, Kaiserstrasse 12, Austria

Equipment Sources

Contact Elko Euzkaldunak Club (*see* above)

Bibliography

a) Webster, David. *The Scottish Highland Games.* Brooklyn: Beekman Publishers Inc.

Index